# STRUCTURE AND BONDING

Volume 24



Editors: J. D. Dunitz, Zürich
P. Hemmerich, Konstanz · R. H. Holm, Cambridge
J. A. Ibers, Evanston · C. K. Jørgensen, Genève
J. B. Neilands, Berkeley · D. Reinen, Marburg
R. J. P. Williams, Oxford

With 40 Figures

Springer-Verlag
Berlin Heidelberg GmbH 1975

ISBN 978-3-662-15514-1 ISBN 978-3-540-37576-0 (eBook)
DOI 10.1007/978-3-540-37576-0

Library of Congress Catalog Card Number 67-11280


Originally published by Springer-Verlag Berlin Heidelberg New York in 1975
Softcover reprint of the hardcover 1st edition 1975

# Contents

STRUCTURE AND BONDING is issued at irregular intervals, according to the material received. With the acceptance for publication of a manuscript, copyright of all countries is vested exclusively in the publisher. Only papers not previously published elsewhere should be submitted. Likewise, the author guarantees against subsequent publication elsewhere. The text should be as clear and concise as possible, the manuscript written on one side of the paper only. Illustrations should be limited to those actually necessary.

---



---


SPRINGER-VERLAG

D-6900 Heidelberg 1
P. O. Box 105280
Telephone (06221) 487·1
Telex 04-61723

D-1000 Berlin 33
Heidelberger Platz 3
Telephone (030) 822001
Telex 01-83319

SPRINGER-VERLAG
NEW YORK INC.

175, Fifth Avenue
New York, N. Y. 10010
Telephone 673-2660

# Photo-electron Spectra of Non-metallic Solids and Consequences for Quantum Chemistry

**C. K. Jørgensen**

Département de Chimie minérale et analytique, Université de Genève
CH-1211 Geneva 4, Switzerland

**Table of Contents**

# I. Chemical Shifts of Inner Shells

## A. Quantum-chemical Remarks about the Continuum

When a sample is irradiated with mono-energetic photons $h\nu$, the ejected electrons show a distribution of kinetic energies $E_{\mathrm{kin}}$. The main reason why photo-electron spectra (the relative number of electrons as a function of $E_{\mathrm{kin}}$) are interesting is that the distribution shows sharp maxima, *signals*, typically showing one-sided half-widths $\delta$ between 0.5 and 3 eV when solids are bombarded with $h\nu = 1253.6$ eV (originating in a magnesium anti-cathode) or 1486.6 eV (from aluminium) whereas the signals may be far narrower, below 0.05 eV, when gases are bombarded with $h\nu = 21.2$ eV (the resonance line of the helium atom) or 40.8 eV ($2p \rightarrow 1s$ of $He^+$). The positions of these maxima are interpreted as *ionization energies I via*

$$I = h\nu - E_{\mathrm{kin}} \tag{1}$$

This interpretation is no conceptual difficulty in the case of the lowest $I$ (by the way, it corresponds to the *Einstein* threshold of photo-electron emission from metallic surfaces, and to *Franck*'s determination of $I$ of gaseous atoms resulting in the name "ionization potential" which is now disappearing) but the interesting point is that $I$ of penultimate MO (molecular orbitals) and of inner shells also can be determined.

That inner shells have verifiable energies, has been recognized since the time of *Moseley* using the *Ritz* combination principle $h\nu = (E_1 - E_2)$ on X-ray emission lines, and in the *Hartree-Fock* calculations (*1*) on monatomic entities, a definite mathematical expression $\varepsilon$ is expected to indicate the observable $I$. Actually, the $I$ values are usually slightly lower (*2*) to the extent 0.8 eV times the square-root of $I$ (in eV). As pointed out by *Koopmans* (*3*) the major part of this discrepancy is due to assumption of "frozen orbitals", that the electron density only varies by the removal of one electron from a definite orbital. If one calculates the *rearrangement energy* by performing a *Hartree-Fock* calculation on the ionized system, allowing the other orbitals to adapt (in monatomic entities, they contract their radial functions) good agreement is obtained with $I$ as the energy difference between the *Hartree-Fock* functions for the ionized and for the original system. The former wave-function is a constrained *Hartree-Fock* function in the sense that definite occupation numbers of the shells are prescribed (one of the inner $nl$-shells contain only $(4l+1)$ electrons) and one must be careful in minimizing the energy using the variation principle not to mix, say $1s^2 2s 2p^6$ of $Ne^+$ with the desired $1s 2s^2 2p^6$. The *correlation energy* is the (negative) difference between the actual energy (corrected for relativistic effects) and the *Hartree-Fock* energy (*4, 5*). Normally, the correlation energy is less negative in the ionized system, and hence, it contributes a small, but negative, quantity to $(\varepsilon - I)$. In monatomic entities, the difference between the rearrangement energy and the small decrease of the correlation energy is called the *intra-atomic relaxation energy*.

However, for chemists, it is fascinating that the penultimate MO also can be detected. It was very difficult previously to connect the eigen-values of an effective one-electron operator in an approximate MO calculation (rarely attaining *Hartree-Fock* accuracy) with experimental results. Hence, it was a great progress that *Turner* and his group *(6)* identified many penultimate MO of diatomic and triatomic molecules in excellent qualitative agreement with the LCAO (linear combination of atomic orbitals) model and group-theoretical arguments *(7)*. However, the order of MO having different symmetry types sometimes shows inversions, because the ratio (usually between 0.8 and 0.95) between $I$ and calculated eigen-values is not exactly the same for different orbitals of the same molecule. The vibrational structure discussed below sometimes produces problems comparable to the ultra-violet spectra of polyatomic molecules *(8)* where the individual electronic transitions are difficult to distinguish. Thus, it is beyond doubt that the 15 orbitals containing the 30 valence electrons of benzene $C_6H_6$ have 10 $I$-values (for group-theoretical reasons, five pairs of MO in the point-group $D_{6h}$ have necessarily coinciding $I$) distributed between 9.3 and 25.8 eV, but the detailed assignment of the photo-electron signals in this region is still discussed *(9, 10)*. One aspect of this controversy is the vibrational structure with several maxima seen in *Jahn-Teller* unstable, ionized hydrides. Thus, methane $CH_4$ has a complicated structure centered around $I = 14$ eV corresponding to removal of an electron from one of the three orbitals of symmetry type $t_2$ degenerate in the point-group $T_d$. The two other signals from methane, obtainable with higher $h\nu$, show $I = 23$ eV of the totally symmetric ($a_1$) orbital known *(11)* to be a linear combination of carbon 2*s* and hydrogen 1*s* orbitals, and $I = 290.8$ eV for the other $a_1$ orbital extremely close to being a pure carbon 1*s* orbital. It is interesting to compare with the 10-electron atom neon having $I(\mathrm{Ne}2p) = 21.6$, $I(\mathrm{Ne}2s) = 48.4$ and $I(\mathrm{Ne}\,1s) = 870.3$ eV, and with the water molecule, where *(6, 12)* $I = 12.6$, 14.8, 18.6, 32.2 and 539.7 eV correspond to four MO having different symmetry types, and the high $I$ essentially oxygen 1*s*. In the point-group $C_{2v}$ of $H_2O$, the three Cartesian axes are non-equivalent, and it would be an argument of "ligand field" theory *(7)* that the three 2*p* orbitals of the oxygen atom participating in the chemical bonding separate in a non-bonding (perpendicular on the molecular plane) and two bonding orbitals.

In gaseous samples, $E_{\mathrm{kin}}$ of the electron liberated by the primary process described in Eq. (1) is not frequently changed by subsequent processes, whereas the major part of electrons from a solid sample have undergone inelastic scattering. The first-order result is that the electrons not participating in the signal peak get distributed rather evenly over all $E_{\mathrm{kin}}$ from the lower limit measured by the instrument (say 100 eV) to the signal, with the result that each signal is accompanied to the left (usually, the photo-electron spectra are recorded with $E_{\mathrm{kin}}$ increasing toward the right and $I$ increasing toward the left) with a roughly constant contribution forming a staircase raising in steps at each new signal towards the left. Hence, it is rather exceptional that a peak is more than twice as high as the mildly undulating background. This is connected with the fact that the typical average escape depth of electrons not suffering inelastic scattering is 20 Å (this value depends to some extent on the sample) hundred times less than the average penetration of soft X-rays.

The primary ionization process follows *Franck* and *Condon*'s principle of unchanged internuclear distances, like visible and ultra-violet spectra *(8)*. Hence, the centre of gravity (*barycentre*) of the vibrational structure frequently seen in photoelectron spectra of gaseous molecules (*6*) corresponds to "*vertical*" $I$ with invariant internuclear distances, whereas the slightly lower $I$ of the origin sometimes can be detected, and is called the *adiabatic ionization energy*. Even when no vibrational structure is resolved, the removal of a strongly bonding (or strongly anti-bonding) electron produces a broadened signal where $\delta$ readily can be 0.5 to 1 eV (like in absorption spectra (*13, 14*) where $\delta$ usually is 0.2 to 0.4 eV for transitions between two MO). It is important to remember that this *Franck-Condon* contribution can be perceptible for ionization of inner shells, such as 1*s* of lithium, nitrogen and oxygen (*15*). Another source of width of signals is short half-life (due to *Auger* transitions or X-ray emission accompanying less strongly bound electrons jumping into the vacancy) where $\delta$ according to *Heisenberg*'s uncertainty principle is 2.3 eV divided by the half-life in the unit $10^{-16}$ sec. Whereas instrumental broadening and *Franck-Condon* broadening (if considered an effect of a harmonic oscillator (*13*) vibration) tend to produce Gaussian error-curves, Heisenberg broadening tend toward the Lorentzian shape = constant $/[(I - I_0)^2 + \delta^2]$. Quite generally, the observed $\delta$ can be considered as a square-root of the sum of the squared individual contributions toward the signal width.

However, photo-electron spectra also follow an electronic analogy to the *Franck-Condon* principle, which may be called the *Manne-Åberg principle* (*16*). These authors pointed out that a neon atom shows satellites at $I$ values between 900 and 1000 eV, above 870.3 eV for the primary signal due to formation of (relaxed) $1s2s^22p^6$ of $Ne^+$. These satellites are either due to *shake-up* forming excited configurations such as $1s2s^22p^53p$ of $Ne^+$ or to *shake-off* loosing simultaneously two or more electrons, forming levels belonging to configurations such as $1s2s2p^6$ of $Ne^{+2}$ or $1s2s2p^5$ of $Ne^{+3}$. This phenomenon (*12*) should not be confused with *Auger transitions*, as when the ionized $1s2s^22p^6$ subsequently decay to $^3P$, $^1D$ or $^1S$ of $1s^22s^22p^4$ of $Ne^{+2}$ or $^3P$ or $^1P$ of $1s^22s2p^5$ of $Ne^{+2}$ under emission of an *Auger* electron. *Manne* and *Åberg* (*16*) pointed out that though these satellites at most have 4 percent each of the intensity of the main peak at $I = 870.3$ eV, their integrated intensity is a quarter of the main peak. The interesting point is that the barycentre of all this structure (of which 80 percent is concentrated on the main peak) occurs at $I = 890$ eV in agreement with the *Koopmans* value of $\varepsilon$. Correspondingly, one can describe the ionization in the "sudden approximation" as the ionized system choosing between the eigenstates of the *new* Hamiltonian because the remnant $1s2s^22p^6$ of the original system has inappropriately extended radial functions. However, the overlap integral between the remnant and the new, adapted $1s2s^22p^6$ is close to 0.9 explaining why 80 percent of the processes result (provisionally) in the eigenstate $1s2s^22p^6$ of $Ne^+$.

It is worthwhile to consider this result from the point of view of the Copenhagen principle of *final states* that the outcome of a process like the primary ionization is distributed on a definite manifold of probabilities (adding up to 100 percent) of realizing each possible final state. Thus, the neon atom functions as an honest gambling-machine; when it has absorbed the 1486.6 eV photon, it has 2 percent probability of forming $1s^22s^22p^5$ (the groundstate of $Ne^+$) producing the Ne2*p* signal with $E_{kin} = 1465.0$ eV; 4 percent of forming $1s^22s2p^6$ (corresponding to the Ne2*s* signal with $E_{kin} = 1438.2$ eV); 76 percent of forming $1s2s^22p^6$ (the main Ne1*s* peak with $E_{kin} = 616.3$ eV); 8 percent of forming various shake-up states, mostly belonging to $1s2s^22p^53p$; and 10 percent of various shake-off states, the highest $E_{kin} = 569$ eV and strongest signal representing $1s2s^22p^5$ of $Ne^{+2}$. The choice of the individual neon atom between these alternatives has no cause in the Medieval sense, any more than the exact time for decay of a radioactive nucleus. It is conceivable (*17*, *18*) that the *Copenhagen* principle of final states does not apply to systems which are so large and complicated that they cannot be reproduced as indiscernible entities; but photo-electron spectrometry has supplied additional evidence for this principle, and about the general time-scale of the measurement which is situated somewhere between $3 \cdot 10^{-18}$ sec (the time of passage of one wave-length of the soft X-rays) and well below $5 \cdot 10^{-17}$ sec, which is the *Heisenberg* half-life of the cadmium $4p$ ionized situation (*15*). Hence, this new technique is very close to give an instantaneous picture (*7*) in contrast to nuclear magnetic and quadrupolar resonance, electron spin magnetic resonance, and *Mössbauer* spectrometry (*19*, *20*) having longer time-scales.

## B. Fractional Atomic Charges

When it is remembered that 1 eV = 23.05 kcal/mole = 8066 $cm^{-1}$ it is astonishing to the chemist that the shift $dI$ of ionization energies of inner shells in different compounds of the same element frequently amounts to 5 eV when it is realized that the dissociation energies of strong single bonds such as $H_2$ is 4.6 eV and of each C—C bond in diamond 3.7 eV. It should not be construed that the total energy is modified to an extent comparable to $dI$. Actually, it may even be argued (*4*) that $dI$ is only an indicator of the changed *Hartree* potential $U(x, y, z)$ averaged over the positions of the orbital to be ionized. The point is that the total energy of a monatomic entity relative to its nucleus and electrons all being highly separated is the negative

$$E_{tot} = -E_{nucl/electr} + E_{kin} + E_{electr/electr} \tag{2}$$

where the three quantities on the right-hand side (nuclear attraction of the electrons, kinetic energy of the electrons, and their mutual repulsion) are written positive. In electrostatic systems, the virial theorem

$$E_{kin} = -E_{tot} = -\frac{1}{2} E_{pot} \tag{3}$$

is valid, with the result that Eq. (2) can be written in purely potential contributions (which obviously cannot be treated by the variation principle, since a stable electrostatic system becomes $K$ times more stable if all distances are decreased by the same factor $K$)

$$2\,E_{tot} = -E_{nucl/electr} + E_{electr/electr} \tag{4}$$

In the approximation (neglecting correlation effects) of a well-defined electron configuration with $q$ electrons in definite orbitals

$$E_{\text{nucl/electr}} = \sum_q Z\langle r^{-1}\rangle$$

$$E_{\text{electr/electr}} = \frac{q(q-1)}{2} \text{ quantities of the form } A_*(a,b)\ . \tag{5}$$

We have here chosen to consider the barycentre (*21*, *22*) of all the states belonging to the configuration, in which case

$$A_*(a,b) = J(a,b) - \frac{1}{2} K(a,b) \tag{6}$$

where $J(a,b)$ is the classical electrostatic interaction between the extended charge distributions $\psi_a^2$ and $\psi_b^2$ and $K(a,b)$ is the two-electron operator exchange integral representing the electrostatic interaction between the product of the two orbitals $\psi_a\psi_b$ and itself. Normally, $K(a,b)$ is much smaller than $J(a,b)$, either because $\psi_a$ and $\psi_b$ are rather separate in space, or because the product $\psi_a\psi_b$ is as frequently negative as it is positive, in the sense that the overlap integral (which is the integral of the product over all space) is zero. In the special case of $a=b$, $J(a,a)$ is the interaction between $\psi_a^2$ and itself, whereas Eq. (6) is written such a way that $K(a,a)$ vanishes. In Eq. (5) and in the following equations, atomic units are used, the charge $e$ of the proton being unity, 1 bohr $=0.529$ Å of length and 1 hartree $=$ 2 rydberg $=27.2$ eV of energy.

The *one-shot ionization energy* of one of the orbitals $a$ is

$$I(a) = Z\langle r^{-1}\rangle - \sum_{k\neq a} A_*(k,a) - E_{\text{kin}}(a) \tag{7}$$

in this approximation, and the sum of the $I(a)$ for all $q$ electrons is

$$E_{\text{nucl/electr}} - 2\,E_{\text{electr/electr}} - E_{\text{kin}} = -\,E_{\text{tot}} - E_{\text{electr/electr}} \tag{8}$$

considerably smaller than the sum of all the *consecutive* ionization energies for all $q$ electrons, being $-E_{\text{tot}}$ by definition. Normally, one concentrates interest on inner shells screening off much of the nuclear attraction (almost to the extent of the apparent value of $Z$ diminuishing one unit for each electron in an inner shell) but it should not be neglected that *external screening* from electrons in orbitals with larger average radii also contribution to decreasing the one-shot ionization energy of an inner shell. Thus, $I$(O 1$s$) of various oxygen compounds vary between 535 and 542 eV, whereas the seventh consecutive $I_7$ of $O^{+6}$ isoelectronic with helium is 739.3 eV and $I_8$ of the last 1$s$ electron 871.4 eV. Actually, $I_6 = 552.1$ eV of $N^{+5}$ showing that the eight valency electrons in the water molecule have the same effect on $I$(O 1$s$) as the removal of one proton from the nucleus would have. It may be noted that $I$(1$s$) from lithium to aluminium is very close to $(Z^2 - Z)$ 9.7 eV, where the constant 9.7 eV is only 0.7 rydberg.

The parameters of interelectronic repulsion $A_*(a,b)$ for two orbitals with highly different average radii are approximately $\langle r^{-1}\rangle$ hartree/bohr (this unit corresponds to 14.4 eV/Å) of the *largest* orbital with the smallest average reciprocal radius. *Politzer* and *Daiker* (*23*) investigated the small deviations from this hypothesis. On the other hand, for two orbitals with identical extension, $A_*$ is between 0.6 and 0.7 (depending on the actual shape) times $\langle r^{-1}\rangle$ in atomic units. Hence, for a given electron, it is a fairly good approximation to say that the $Z_c$ electrons in core orbitals with distinctly smaller radii decrease the effective nuclear charge felt by the electron to $(Z - Z_c)$ but the *screening constant* $Z_s$ frequently used by X-ray spectroscopists identifying $I = (Z - Z_s)^2$ rydberg$/n^2$ (where $n$ is the principal quantum number) is larger than $Z_c$ because it contains further contributions from electrons in the same shell as the electron to be ionized, and from external screening. The approximate contribution to $Z_s$ from each external electron is $\langle r^{-1}\rangle_e$ of the external electron divided by $\langle r^{-1}\rangle$ for the electron considered.

It may be worthwhile to write the *Schrödinger* equation for one electron moving in the potential $U(x,y,z)$ having the (negative) eigen-value $E$ for stationary states with the real wave-function $\psi$

$$U(x,y,z) + \mathscr{P}(x,y,z) = E \tag{9}$$

written on the conservative form with the local operator of kinetic energy

$$\mathscr{P} = -\frac{1}{2\psi}\left[\frac{\partial^2\psi}{\partial x^2} + \frac{\partial^2\psi}{\partial y^2} + \frac{\partial^2\psi}{\partial z^2}\right] \tag{10}$$

which is negative in regions with $E < U$ (in contrast to the non-negative $\frac{1}{2}\,\mathrm{mv}^2$ in classical mechanics). $E_{\mathrm{pot}}\ (= 2E)$ is the integral $\int U\psi^2 d\tau$ over all spatial variables which would normally be written $\langle U\rangle$, and $E_{\mathrm{kin}} = \int \mathscr{P}\psi^2 d\tau = \langle\mathscr{P}\rangle$. One should not be worried about the division with $\psi$ in Eq. (10), $\mathscr{P}$ only diverges towards $+\infty$ at the singularities (the nuclei considered as points) where $U \to -\infty$. *Ruedenberg* (*24*) discusses why it is not against *Heisenberg*'s uncertainty principle to select three Cartesian coordinate directions and evaluate the three contributions $\langle\mathscr{P}_x\rangle$, $\langle\mathscr{P}_y\rangle$ and $\langle\mathscr{P}_z\rangle$ in Eq. (10). It is possible to discuss other operators of kinetic energy (*21*) and the question (*7*) to what extent *Hellmann* was right in suggesting that the chemical bonding is connected with the local contribution to the kinetic energy in the bond region.

In a monatomic entity, the nucleus can be put in origo, and the potential $U(r)$ depends only on $r$ (the technical word is "central field" in spherical symmetry) and is $-Z/r$ for one-electron systems. The one-electron functions $\psi$ in spherical symmetry have the property to be products of an *angular function* $A_l$ and a *radial function* $R_{nl}$

$$\psi = A_l \cdot R_{nl} \,/\, (2\sqrt{\pi}\, r) \tag{11}$$

which we have chosen to normalize in such a way that $(A_l)^2$ on the average is 1 on a spherical surface with the differential volume $4\pi r^2 dr$, and the integral of $(R_{nl})^2$ from $r = 0$ to infinity is also 1. For a given value of $l$, one can select (*7, 21*) $(2l + 1)$ mutually orthogonal $A_l$ which are each a linear combination of homogeneous polynomials $(x^a y^b z^c/r^l)$ in the three *Cartesian* coordinates with the sum $a + b + c$ of the three non-negative integers $(a, b, c)$ equal to $l$. The radial function $R_{nl}$ close to the nucleus is proportional (with a numerical constant depending on the behaviour in the whole $r$ interval) to $r^{l+1}$ and it posses $(n - l - 1)$ radial nodes, *i.e.* positive, finite $r$ values for which $R_{nl} = 0$ (and changes sign) whereas $A_l$ has $l$ angular nodes (a node-cone counts for two node-planes). The presence of radial or angular nodes amplify the kinetic energy according to Eq. (10). In the special case of spherical symmetry, the operator $\mathscr{P}$ can be divided in an angular and a radial part

$$\mathscr{P}_{\text{ang}} = l(l+1) \,/\, 2r^2$$
$$\mathscr{P}_{\text{rad}} = - \frac{1}{2R_{nl}} \frac{\partial^2 R_{nl}}{\partial r^2} \tag{12}$$

the former having the unusual property of explicit dependence on $r$ without reference to other aspects of $\psi$ than the quantum-number $l$. Hence, $\mathscr{P}_{\text{ang}}$ is frequently called the *centrifugal pseudo-potential* and added to $U$ when the *Schrödinger* equation is solved.

The so-called *hydrogenic radial functions* are the $R_{nl}$ in H, $He^+$, $Li^{+2}$, ... containing one electron. It is possible (*25*) to evaluate average values (in $\text{bohr}^n$) of $r^n$ for such $(R_{nl})^2$ of importance for the following discussions

$$\langle r^2 \rangle = n^2(5\,n^2 - 3\,l(l+1) + 1) \,/\, 2\,Z^2$$
$$\langle r \rangle = (3\,n^2 - l(l+1)) \,/\, 2\,Z$$
$$\langle r^{-1} \rangle = Z/n^2 \tag{13}$$
$$\langle r^{-2} \rangle = Z^2 \,/\, n^3\left(l + \frac{1}{2}\right)$$
$$\langle r^{-3} \rangle = Z^3 \,/\, n^3(l+1)\left(l + \frac{1}{2}\right) l$$

The most significant is the reciprocal radius serving to know $E_{\text{pot}} = - Z\langle r^{-1} \rangle = - Z^2/n^2 = 2E$, but is interesting to note that

$$E_{\text{ang kin}} = Z^2\, l(l+1) \,/\, 2\,n^3\left(l + \frac{1}{2}\right)$$
$$E_{\text{rad kin}} = Z^2 \left\{ n\left(l + \frac{1}{2}\right) - l(l+1) \right\} \Big/ \; 2\,n^3\left(l + \frac{1}{2}\right) \tag{14}$$

with the result that $E_{\text{ang kin}} \,/\, E_{\text{rad kin}}$ for *hydrogenic* radial functions is

| | | | |
|---|---|---|---|
| $2p$: 2 | $3d$: 4 | $4f$: 6 | |
| $3p$: $^4/_5$ | $4d$: $^3/_2$ | $5f$: $^{24}/_{11}$ | (15) |
| $4p$: $^1/_2$ | $5d$: $^{12}/_{13}$ | $6f$: $^4/_3$ | |

The whole structure of the Periodical Table is connected with the competition between (high $n$, low $l$) and (lower $n$, higher $l$) shells in a central field $U(r)$ having the property that the product $rU$ smoothly decreases from $Z$ at the nucleus (the external screening increases only $U$ by a constant at origo) to $(z+1)$ at large $r$, where $z$ is the ionic charge. It is easy to show qualitatively (*21*) that such a central field is less tolerant to high $l$-values than the hydrogenic (Coulombic) case $rU = Z = (z+1)$. It is an empirical fact (*26, 27*) that the gaseous ions $M^{+2}$, $M^{+3}$, $M^{+4}$, ... follow the *Aufbau principle* by having groundstates belonging to the configuration obtaining by allocating the $(Z-z)$ electrons to the shells in the order

$$1s \ll 2s < 2p \ll 3s < 3p \ll 3d < 4s < 4p \ll 4d < 5s < 5p \ll \\ 4f < 5d < 6s < 6p \ll 5f < 6d < 7s \ldots \quad (16)$$

where the double inequality signs indicate the configurations [He], [Ne], ... isoelectronic with the noble gases. The five known exceptions to the series (*16*) all are $M^{+2}$ with M = lanthanum [Xe]$5d$, gadolinium [Xe]$4f^7 5d$, lutetium [Xe]$4f^{14}6s$, actinium [Rn]$7s$ and thorium [Rn] $5f\,6d$ whereas the groundstate of $Th^{+3}$ belongs to [Rn]$5f$. By a mistake, it was previously stated (*28*) that only four exceptions occur. Another addendum to this review (*28*) of the lanthanides is that *Stoner* (*29*) already in 1924 was the first to suggest shells each able to accommodate $(4l+2)$ electrons, rationalizing the interval $Z = 57$ to 71.

It is of utmost importance to chemistry that the preponderant configurations determining the oxidation states of transition group compounds (*7*) also follow the Aufbau principle (*16*) in negative oxidation numbers as well as M(0) and M(I) in sharp contrast to the neutral atoms preferring (with exception of the palladium atom [Kr]$4d^{10}$) to have one or two $(n+1)s$ electrons replacing one or two of the $nd$. Thus, the complexes $Fe(CO)_4^{-2}$, $Co(CO)_4^{-}$, $Co(PF_3)_4^{-}$, $Ni(CO)_4$ and $Ni(PCl_3)_4$ containing Fe(−II), Co(−I) and Ni(0) are all closed-shell [Ar]$3d^{10}$ systems isoelectronic with Cu(I), Zn(II), Ga(III), ... (of which the photo-electron spectra (*30*) have been discussed) whereas the groundstate of the gaseous nickel atom belongs to [Ar]$3d^8 4s^2$ (though the lowest configuration barycentre is [Ar]$3d^9 4s$) as $Co^-$, and $Ni^-$ [Ar]$3d^9 4s^2$. Gaseous $Fe^{-2}$ would spontaneously loose an electron like $O^{-2}$ and all other $M^{-2}$.

The $(n+1)s$ electrons in neutral transition-group atoms are probably replaced by the electrons in MO forming covalent bonds in condensed matter (solution, amorphous and vitreous materials, stoichiometric or substituted crystals) or by the conduction electrons in the metals, as discussed by *Watson* and *Perlman* in this volume. The main rôle of such $s$-like electrons is to supply the external screening needed for stabilizing a definite number of electrons in the partly filled shell. In the approximation Eq. (5) configurations such as $d^m s^n$ have the energy

$$-\,m\,I_\varepsilon(d) - n\,I_\varepsilon(s) + \frac{m(m-1)}{2}A_*(d,d) + mn\,A_*(s,d) + \frac{n(n-1)}{2}A_*(s,s) \quad (17)$$

relative to the previous noble-gas configuration with $(m+n)$ fewer electrons, where $I_\varepsilon$ is the ionization energy of a single $d$ or of a single $s$ electron added to this noble-gas configuration. In the typical case (*2, 22*) of $I_\varepsilon(d) > I_\varepsilon(s)$ and $A_*(d,d) > A_*(s,d) \sim A_*(s,s)$, Eq. (17) predicts a parabolic (and not linear, as would be true for $2A_*(s,d) = A_*(d,d) + A_*(s,s)$) dependence of the configuration barycentre energy as a function of the number $n$ of $s$ electrons, keeping $(m+n)$

and hence $z$ constant. Actually, the barycentre of [Ar]$3d4s$ in $Sc^+$ and of [Ar]$3d^94s$ of gaseous Ni constitute rather accurately (*37*) the minimum of the parabola, the two other barycentres with zero or two $4s$ electrons being situated at comparable, higher energies. Whereas $s$ electrons begin to be stabilized for relativistic reasons in atoms with $Z$ above 70, the $s$ orbital is always anti-bonding in compounds, explaining the chemical behaviour and distorted symmetry (corresponding to mixing of $6s$ and $6p$ orbitals) in the *Gillespie*- unstable(*7, 32*) thallium(I), lead(II) and bismuth(III).

It has been recognized for many years (*13, 33*) that the fractional atomic charge of atoms containing partly filled $d$ shells is considerably smaller than the oxidation number, though evidence from the parameters of interelectronic repulsion (the *nephelauxetic*, or cloud-expanding, *effect*) normally suggest fractional charges above +1 (*21, 34*) in disagreement with the "electroneutrality principle" proposed by *Pauling* that such charges always are between −1 and +1.

It has never been shown that the chemical shifts $dI$ for inner shells depend on the oxidation state only. In gaseous compounds (free of the disturbing charge effects discussed in the next chapter) a strong influence of the electronegativity of the neighbour atoms is perceptible. Thus (*35*), $I$(N 1$s$) = 417.0 eV of $ONF_3$, 414.2 $NF_3$, 412.5 (central) and 408.6 (terminal) NNO, 409.9 $N_2$, 406.8 HCN, 405.6 $NH_3$ and 404.7 eV of $N(CH_3)_3$ and in boron compounds (*36*), $I$(B 1$s$) = 202.8 eV of $BF_3$, 199.8 $BCl_3$, 199.0 $BBr_3$, 197.8 $BI_3$ and $B(OCH_3)_3$, 196.5 $B_2H_6$, 195.5 $B(C_2H_5)_3$ and 193.2 eV in the adduct $H_3BN(C_2H_5)_3$ to be compared with values (*15*) close to 201 eV for salts of $BF_4^-$ and 193 eV for tetraphenylborates containing $B(C_6H_5)_4^-$. The only elements for which a clear-cut dependence on oxidation state subsists, are the lanthanides (*15, 28*) showing $dI$ for inner shells between 12 and 16 eV when comparing Ce(III) with Ce(IV), Eu(II) with Eu(III) or Tb(III) with Tb(IV). The corresponding change of I(4$f$) is 9 eV between isoelectronic $4f^q$ systems. The other transition groups, and in particular uranium compounds (*15, 37*), do not show such dramatic effects of changing the oxidation state. For a given element in a given oxidation state, the variation of $I$-values is typically 2 to 5 eV (*15*). In the case of transition and post-transition elements, the highest $I$ are found for fluorides and the lowest for oxides, sulphides and complexes of sulphur-containing ligands. This situation that the neighbour atoms are at least as important as the oxidation state is completely extreme (*15*) in $Tl_2O_3$ having lower $I$ than all 21 thallium(I) compounds measured, and $PbO_2$ lower I than all 15 lead(II) compounds. Further on, most Co(III) have lower $I$ than Co(II). The first case reported of such inversions (*38*) is the study by *Kramer* and *Klein* finding $I$(Fe3$p$) varying $FeO_4^{-2} \sim FeF_6^{-3} > Fe(H_2O)_6^{+3} > Fe(H_2O_6^{+2} > Fe(CO)_5 >$ iron(III) complexes of sulphur-containing ligands.

There was general agreement that the major origin of these shifts is the variation of the fractional atomic charge, though already in 1968 the importance of the *Madelung* potential (to be discussed below) was emphasized (*39*). The theoretical estimate of $dI$ per unit of atomic charge is rather large and has the order of magnitude of $\langle r^{-1}\rangle$ atomic units of the orbital changing the occupation number when $z$ changes. This quantity is about 40 eV for $4f$ group $M^{+3}$, about thrice the shift observed, but also 20 to 30 eV for $3d$ group $M^{+2}$ and $M^{+3}$, much larger than the typical shifts 0 to 6 eV observed between M(II) and M(III) compounds. For the elements from boron to fluorine, the $2p$ radial function is roughly hydrogenic with the *Slater* exponent $\mu$ varying from 1.15 to 2.45, and here, $\langle r^{-1}\rangle = \mu \cdot 27.2$ eV. The discrepancy can be discussed along two lines. Either, one is inspired by the electroneutrality principle and assumes fractional charges between 0.1 and 0.3 multiplied by the theoretical value of $\langle r^{-1}\rangle$ or one may investigate large compensating effects related to the *Madelung* potentials and the relaxation energy. Many authors have attempted comparison between fractional atomic charges evaluated for LCAO approximations, especially of carbon, nitrogen (*12, 40*) and sulphur (*12*) compounds. Though the agreement sometimes is surprisingly good, it cannot be excluded that the slope of the linear correlation is wrong by a factor of 2 or 4. *Stucky, Matthews, Hedman, Klasson* and *Nordling* (*41*) made a careful comparison of such data (including the *Madelung* potential) and found the slopes (per unit $z$) 20.7 eV for $I$(C 1$s$) and 31 eV for $I$(N 1$s$). The zero-point for a neutral carbon atom in a gaseous molecule is 291.7 eV (to be compared with $I = 301.7$ eV in $CF_4$, 296.2 in $CCl_4$ and 290.8 eV in $CH_4$). It may be noted that the fractional charges in this treatment (the extreme being $+1.0$ and $-0.25$ in $CF_4$) are smaller than obtained (*21*) in the model of differential ionization energies. However, it is very difficult at present to give a significant value for the fractional charge of the atoms in the closed-shell molecules $CF_4$. Recently, xenon fluorides (*270*) and 144 other gaseous molecules (*271*) have been treated along these lines.

### C. Madelung Potentials and the Charging Effect

Already before 1930, it attracted attention (*42*) that the ionization energies $I_z$ going from $M^{+z-1}$ to $M^{+z}$ in the gaseous state tend to be proportional to $z$ (and the total energy relative to the neutral atom hence proportional to $(z^2+z)$ as long one does not cross a closed shell) and that they are overcompensated by the chemical bonding to the extent that certain cases, such as beryllium(II), aluminium(III), erbium(III) and thorium(IV) are the only well-known oxidation state in non-metallic compounds of the element considered. Three of the examples given are, like all the alkaline and alkaline-earth elements, instances of crossing shells (the electron affinity $I_3 = 28.45$ eV of $Al^{+3}$ acquiring a $3s$ electron is far smaller than $I_4 = 120.0$ eV loosing one of the $2p$ electrons from a closed shell). The situation is more subtle in the case of $Er^{+3}$ also having (*43*) a larger $I_4 = 42.6$ eV going from the groundstate of $4f^{11}$ to $4f^{10}$ than the electron affinity $I_3 = 22.7$ eV corresponding to the energy difference between the lowest levels of $4f^{11}$ and $4f^{12}$. Hence, the difference $(I_4 - I_3) = 19.9$ eV is large because of the small average radii of the partly filled $4f$ shell, and is not of the same origin as $I_3 = 28.0$ eV and $I_4 = 54$ eV of indium being an attenuated example of the closed shells of $Al^{+3}$.

In the following two decades, most chemists were convinced about the validity of the electroneutrality principle and were no longer surprised by this compensation of ionization energies by chemical bonding. However, by the same token as the existence of homonuclear molecules (such as $H_2$, $N_2$, $P_4$ and $S_8$) clearly show that bonding is not exclusively electrostatic, it was also admitted that some salts of univalent anions of the type NaCl and $CaF_2$ are ionic though (legitimate) doubts were expressed about the isotypic chalkogenides MnO, green MnS, NiO, CdO, $CeO_2$ and $ThO_2$.

In a large molecule or in a crystal, one has to lump the three contributions $-E_{\text{nucl/electr}} + E_{\text{electr/electr}}$ from Eq. (2) and the nuclear interrepulsion ($+E_{\text{nucl/nucl}}$) together in a *Hartree* field $U(x,y,z)$. The point is that each of these three quantities diverge (*7*) by being proportional to the fifth power of the linear extension of samples with the same shape and composition, whereas the total energy only varies as the third power. In a small molecule (*21*) it is easy to calculate the *Madelung potential* $V(x,y,z)$ in a given point as the sum of $-q_i/R_i$ of the other atoms $i$ carrying charges $q_i$ at a distance $R_i$ from the point considered. In the case of spherically symmetric ions (such as ions isoelectronic with the noble gases, or half-filled shells such as $S = 5/2$ $3d^5$ Mn(II) and Fe(III) or $4f^7$ Eu(II), Gd(III) and Tb(IV)) the *Madelung* contribution to the total energy can be calculated as if the ions were charged points, on the condition that the ionic charge distributions do not overlap. Thus, the *Madelung* potential of regular tetrahedral $MX_4$, with the percentage ionicity $\xi$ (the fractional charge of M being $+4\xi$ and of each X at the distance $R$ being $-\xi$) at the M site is $+4\xi/R$ and at the $X_{\text{sites}}$ $-2.163\,\xi/R$ with the result that the Madelung energy is $-12.326\,\xi^2/R$. For comparison, quadratic $MX_4$ with all five nuclei in a plane has the potentials $+4\,\xi/R$ and $-2.086\,\xi/R$ producing the Madelung energy $-12.172\,\xi^2/R$ showing a mild electrostatic preference for regular tetrahedral coordination. *E. Madelung* calculated the apparently divergent energies for various crystal lattices. Thus, the *Madelung constant* $\alpha$ corresponds to the potential $+\alpha\xi/R$ at the M site and $-\alpha\xi/R$ at the X site in the cubic types NaCl($\alpha = 1.748$), CsCl($\alpha = 1.763$), CuCl (or zincblende) ($\alpha = 1.638$) and $CaF_2$($\alpha = 5.039$). If the typical compounds are fully ionic, $\xi = 1$, whereas $\xi = 2$ in MnO and $ThO_2$ if they are fully ionic. The *Madelung* energies of these cubic crystals are simply $-\alpha\xi^2/R$ which can be related to the unit cell parameters $a_0$ with the new constant $\alpha^*$ (in the unit eV/Å)

$$\text{NaCl: } 50.2 \qquad \text{CsCl: } 29.3 \qquad \text{CuCl: } 54.4 \qquad CaF_2\text{: } 83.6 \tag{18}$$

used for obtaining the *Madelung* energy $-\alpha^*\xi^2/a_0$.

Even in the case of spherically symmetric, non-overlapping ions, it is not exactly true that the *Madelung* potential $V(x,y,z)$ remains constant. In particular, there is a weak angular dependence having no influence on the energy of spherically symmetric ions, but showing slightly different perturbations on different of the $(2l+1)$ orbitals contributing to differing energy levels of a partly filled shell. The original electrostatic model of the "crystal field" (*44, 45*) considered the (rather small) difference $V(x,y,z) - V_0(r)$ from the spherically symmetric average of the *Madelung* potential the source of separations of energy levels in chromophores $MX_N$ compared with the isolated ion $M^{+z}$. This approach has been abandoned today, as discussed in Chapter IV B.

If we concentrate interest on solid halides, it is quite striking that $I$ of the loosest bound $np$ orbitals of $X^-$ are considerably higher than $I_0$ (in eV) of the gaseous anions (*46, 47*) though not as high as $I_1$ of the neutral atoms:

$$\begin{array}{lllll} I_0: & F^-\ 3.40 & Cl^-\ 3.61 & Br^-\ 3.36 & I^-\ 3.06 \\ I_1: & F\ 17.42 & Cl\ 12.97 & Br\ 11.81 & I\ 10.45 \end{array} \tag{19}$$

An obvious explanation is the *Madelung* potential increasing $I(\mathrm{X}\ np)$ to the extent $+\alpha^*/a_0$ varying in the alkaline halides from 12.5 eV in LiF to 6.4 eV in CsI. The difference between $I_0 = 0.75$ eV and $I_1 = 13.60$ eV for hydrogen is even more pronounced than in Eq. (19) but one cannot hope to detect the very weak H 1*s* signal of hydrides. A corollary of the *Madelung* hypothesis is that the loosest bound *np* orbitals of $M^+$ or $M^{+2}$ have their *I decreased* to the same numerical extent $-\alpha^*/a_0$ as can be confirmed by comparison with known $I_2$ or $I_3$ from atomic spectra (*26, 27*). Several authors (*15, 48–52*) have obtained good agreement with this hypothesis and have also pointed out that the distance between *I* for the penultimate *ns* orbital of the cation and $I(np)$ also remains constant, as one expects from $V(x,y,z)$ adding an exterior perturbation to the central field $U(r)$ effectively producing the new central field $U(r) + V_0(r)$ shifted by a constant. Table 1 gives the $I'$ values for the loosest bound *np* shell of cations and of anions in cubic, fairly electrovalent crystals, where the prime on *I* indicates that correction for charging effects has been made according to the method discussed below.

Contrary to some of the *I* values originally evaluated (*48*) without corrections for the charging effect, it must be admitted that $I'(\mathrm{X}np)$ tends to be lower than $I_0$ from Eq. (19) added to $\alpha^*/a_0$. However, this was also true for the estimate of *Mott* and *Gurney* (*53*) $I = 10.17$ eV of NaCl, 9.49 eV for KCl and 9.00 eV for KBr obtained by using the experimental heat of formation of the crystal from gaseous $M^+$ and $X^-$ rather than the theoretical *Madelung* energy. This approach includes, in a way, the *Born-Fajans-Haber* core-core repulsion, though it might also be argued that the relaxation effects discussed in the next Chapter also contribute to decreasing $I'$. It has even been possible (*54*) to vary the *Madelung* potential by thermal expansion of LiF and KCl decreasing $(1/a_0)$ to a small extent. An important aspect of this experiment is the differences such as $I(\mathrm{Li\,1}s) - I(\mathrm{F\,2}s)$ containing the differential change of $2\alpha^*/a_0$ without any interference from the charging effect.

Since the pioneer work of *Hnatowich, Hudis, Perlman* and *Ragaini* (*55*) it has been clear that finely divided metals such a gold, palladium, and metallic compounds (*15*) such as CuS and $Tl_2O_3$ can be isolated by non-conducting powders with the result that the $I^*$ values recorded by the instrument (relative to a common *Fermi* level some 4 to 5 eV below zero energy of electrons in *vacuo*) are increased some 1 to 4 eV. It is probable that this increase of $I^*$ corresponds to the quasi-stationary potential *V* obtained by the insulator loosing photo-electrons by the combined bombardment of X-rays and some electrons. Monochromatized photons having lost their accompanying electrons are known to produce large, and erratically varying, *V* values. This argumentation may be less certain in the case of mixed non-conductors together with a metallic powder (*15*). Anyhow, it is almost certain that one cannot electrically connect obstinate insulators to the photo-electron spectrometer in spite of the fact that one can isolate conductors on the sample holder. The obvious attempts of evaporating thin layers of gold (*56, 57*) or mixing the sample with powdered graphite, have not been generally successful for eliminating this *charging effect*.

In the writer's opinion, this problem can only be solved to a reasonable approximation by comparing with a standard substance acquiring the same *V* as the non-conductor under study. The most practical method is to distribute the pow-

Table 1. The ionization energy $I'$ of the loosest bound orbitals of the cation M and anion X corrected for the charging effect $\partial$, and the values $I_{calc}$ derived from the monatomic ion and the full Madelung potential in the cubic crystals, all in eV

| | $\partial$ | $I'$(M) | $I_{calc}$ | $I'$(X) | $I_{calc}$ |
|---|---|---|---|---|---|
| LiF | 3.7 | 60.7 | 75.64 − 12.52 = 63.12 | 12.8 | 3.40 + 12.52 = 15.92 |
| NaF | 4.2 | 35.5 | 47.29 − 10.86 = 36.43 | 12.8 | 3.40 + 10.86 = 14.26 |
| NaCl | 2.2 | 35.6 | 47.29 − 8.92 = 38.37 | 10.0 | 3.61 + 8.92 = 12.53 |
| NaBr | 2.4 | 35.3 | 47.29 − 8.45 = 38.84 | 9.0 | 3.36 + 8.45 = 11.81 |
| NaI | 2.8 | 35.6 | 47.29 − 7.77 = 39.52 | 8.4 | 3.06 + 7.77 = 10.83 |
| KF | 2.4 | 21.7 | 31.62 − 9.42 = 22.20 | 11.5 | 3.40 + 9.42 = 12.82 |
| KCl | 2.5 | 21.4 | 31.62 − 7.99 = 23.63 | 9.4 | 3.61 + 7.99 = 11.60 |
| KBr | 1.5 | 21.8 | 31.62 − 7.64 = 23.98 | 9.2 | 3.36 + 7.64 = 11.00 |
| KI | 3.1 | 21.7 | 31.62 − 7.11 = 24.51 | 7.9 | 3.06 + 7.11 = 10.17 |
| RbCl | 2.4 | 19.0 | 27.28 − 7.64 = 19.64 | 9.5 | 3.61 + 7.64 = 11.25 |
| RbBr | 2.4 | 18.5 | 27.28 − 7.31 = 19.97 | 8.9 | 3.36 + 7.31 = 10.67 |
| RbI | 1.7 | 18.8 | 27.28 − 6.85 = 20.43 | 8.3 | 3.06 + 6.85 = 9.91 |
| CsF | 2.5 | 15.9 | 25.1 − 8.35 = 16.75 | 9.8 | 3.40 + 8.35 = 11.75 |
| CsCl | 4.5 | 15.0 | 25.1 − 7.13 = 18.0 | 9.0 | 3.61 + 7.13 = 10.74 |
| CsBr | 3.3 | 15.0 | 25.1 − 6.83 = 18.3 | 8.7 | 3.36 + 6.83 = 10.19 |
| CsI | 3.6 | 15.1 | 25.1 − 6.41 = 18.7 | 8.1 | 3.06 + 6.41 = 9.47 |
| MgO | 3.6 | 54.1 | 80.14 − 23.90 = 56.24 | 10 | |
| $CaF_2$ | 3.2 | 30.6 | 50.91 − 15.35 = 35.56 | 13.3 | 3.40 + 15.35 = 18.75 |
| $SrF_2$ | 3.8 | 25.0 | 43.6 − 14.30 = 29.3 | 13.3 | 3.40 + 14.30 = 17.70 |
| $BaF_2$ | 4.3 | 19.3 | 41.5 − 13.53 = 28.0 | 11.9 | 3.40 + 13.53 = 16.93 |
| ZnS | 2.4 | 15.6 | 39.72 − 20.0 = 19.7 | — | |
| CdO | 2.6 | 16.8 | 37.48 − 21.44 = 16.04 | — | |
| CdS | 1.5 | 16.8 | 37.48 − 18.7 = 18.78 | — | |
| AgBr | 1.9 | 10.0 | 21.49 − 6.72 = 14.77 | — | |
| $CdF_2$ | 1.8 | 17.4 | 37.48 − 15.5 = 22.0 | 11.9 | 3.40 + 15.5 = 18.9 |
| $HgF_2$ | 2.0 | 14.3 | 34.2 − 15.1 = 19.1 | — | |

dered sample on the adhesive side of one-sided scotch tape (we use no. P 600 from the 3M company; their double-sided scotch tape contains C—O—C bridges, and other brands C—Cl bonds) and to use (*48*) the carbon 1*s* signal as internal standard. Since gaseous hydrocarbons (*58, 59*) have $I$(C 1*s*) distributed in a narrow interval between 290.8 and 290.2 eV when the carbon atom considered is surrounded exclusively by hydrogen and carbon atoms, and since the relaxation effects seem to saturate (because $I$ still is 290.3 eV in the larger molecule cyclohexane $C_6H_{12}$) it was suggested (*48*) to put the scotch tape signal at 290.0 eV. This is rather equivalent to the suggestions by previous authors to put the rather ubiquitous

C 1*s* signal on inorganic substances to $I^* = 285$ eV. There are reasons to believe that this signal does not originate from traces of pump oil, but rather from hydrocarbons arriving on the sample from automobile and fuel burner emissions.

However, *Bremser* and *Linnemann* (*60*) pointed out that typical insulators (such as LiF) show two C 1*s* signals, one due to scotch tape hydrocarbon not covered with the powder, and another due to minute amounts of hydrocarbon in so intimate contact with the non-conductor that it has acquired the same positive potential $V$. A closer analysis (*15*) shows that the distance $\partial$ between the two C 1*s* signals can be identified with $V$, and it is argued that $V$ is essentially the distance from the bottom of the empty conduction band of the sample to the Fermi level of the metallic surroundings of the spectrometer. $\partial$ is at most 5 eV, in certain anhydrous fluorides such as $MgF_2$ and $LuF_3$. It is probable that these crystals lack positive electron affinity (conduction band with negative energies) like the case of solid neon, whereas solid xenon and most of the alkaline halides have the electron affinity close to 1 eV. The technique (*15*) is now to define $C'_{st}$ as the difference between 290 eV and $I^*$ recorded by the apparatus for the left of the two carbon signals (corresponding to the hydrocarbon having the potential $V$) and then define $I'$ (hopefully representing the ionization energy of the various shells relative to *vacuo*) as the sum of $C'_{st}$ and the $I^*$ recorded of the various signals. In the case of conducting samples, only one C 1*s* signal is seen (excepting the presence of alcohols, ketones, halocarbons, ...) and $C_{st}$ is the difference between 290 eV and the unique (or right-hand) $I^*$(C 1*s*) producing $I = I^* + C_{st}$ for the other signals. Certain samples become sufficiently conducting under the influence of X-rays to avoid this charging effect $V$; this is particularly true for conjugated constituents such as tetraphenylborates or salts of the methylene blue or tetraphenylarsonium cations, but salts with strong hydrogen bonds such as certain ammonium salts or salt hydrates (not necessarily establishing equilibrium within some hours (*61*) by loosing water at the low pressure $10^{-6}$ torr in the instrument) also tend to show small or vanishing $\partial$. On the whole, less pure samples tend toward lower $\partial$ and correspondingly lower $I^*$ but invariant $I'$. Actually, the samples with measurable $\partial$ between 1.5 and 5 eV have $I'$ reproducible with a precision of 0.1 to 0.2 eV. However, this precision should not be confused with an absolute accuracy which may not be better than 0.5 eV. The largest uncertainty of $I'$ occurs for mixtures of several non-conductors (this can be a considerable obstacle for semi-quantitative analytical applications) and for samples having $\partial$ somewhere between 0 and 1.5 eV. The latter situation can be diagnosized by broadening and asymmetry of the carbon signal.

Another standard is metallic gold (very popular because of the absence of oxidation) where $I^*(\mathrm{Au}4f_{7/2}) = 83.8$ eV (*62*). If this value refers to the *Fermi* level (*63*) $I = 89.1$ eV. Correspondingly, the two standards (gold and oil) are compatible if the distance between the two $I^*$ measured is 200.9 eV. This is indeed the case, but the observed distances do not form a *Gaussian* error-curve around this value, but are distributed with a narrow peak at 200.85 eV and a-quarter of the values deviating 0.3 to 0.6 eV. An argument for the identification of $\partial$ with $V$ is that a sandwich structure of a metallic sample holder, a gold foil in electric contact with the instrument, a piece of scotch tape containing a smaller (isolated) piece of gold foil again partly covered with a second piece of scotch tape have the same distance (slowly decreasing from 3.2 to 2.9 eV during some hours of irradiation) between the two C 1*s* and the two sets of gold (Au $4f_{5/2}$ and Au $4f_{7/2}$ separated *(62)* 3.67 eV) signals *(64)*.

This evaluation of charging effects is related to the *Madelung* potentials since one might have expected $V(x,y,z)$ not to have its regular, extreme value on sites on surfaces of broken crystallites. A shift of 1 to 2 eV would cause a strong asymmetry of the photo-electron signals, which are not at all observed. It is true that stretching the escape-width 20 Å a bit, one may argue that only 10 percent of the ions are directly situated in the surface layer, and that a unconspicuous broadening of cation signals toward the left might remain undetected in the mildly undulating background. However, anion signals should tail off toward the right, which is not evident on the spectra. Hence, there is some compensation mechanism operating, producing essentially the $I'$ values characterizing the bulk of the compounds. Seen from this point of view, the one-sided half-width $\delta(-)$ toward lower $I$ is an important parameter indicating a higher limit to the inhomogeneity of the combined *Madelung* potential and charging effect $V$. Spectra of metallic sodium, magnesium and aluminium measured under conditions of ultra-high vacuo *(65)* show 2*p* signals with an intrinsic half-width $\delta$ below 0.1 eV and 2*s* between 0.15 and 0.35 eV whereas salts of these three elements have intrinsic $\delta$ between 0.25 (Na 2*p* in NaCl) and 1.0 eV (Al 2*s* in $AlF_3$). It was argued *(65)* that the shortening of the half-life due to an *Auger* process involving two final holes in the conduction band of the metal is much less pronounced than the *Auger* process involving two final holes in the 2*p* or 3*p* orbitals of the adjacent anions in the salts. However, there is little doubt that the major part of this broadening originates from the *Franck-Condon* principle, but at the same time, these experiments produce valuable higher limits to the inhomogeneity of the *Madelung* potential.

Using the high-intensity 1253.6 eV source on our Varian IEE-15 photo-electron spectrometer, we find *(66)* $\delta(-) = 0.6$ eV for Ag $3d_{5/2}$ of metallic silver and 0.65 eV for Au $4f_{7/2}$ of gold. It is interesting to note that typical sharp signals of non-conducting compounds have $\delta(-)$ between 0.8 and 1.0 eV, whereas the left C 1*s* signal is close to 0.9 eV and the right close to 0.85 eV demonstrating a remarkable homogeneity of the quasi-stationary potential of the hydrocarbon in intimate electric contact with the sample.

## D. Relaxation Effects

The *Koopmans* value for $I$(C 1*s*) of methane *(67, 68)* assuming "frozen" *Hartree-Fock* orbitals is 305.0 eV, 14.2 eV above the experimental value. Like for atomic neon *(16)* this discrepancy can be remedied within a half eV by performing two *Hartree-Fock* calculations, one for the ionized system constrained to lack one 1*s* electron, and one for the groundstate. *Aarons, Guest, Hall* and *Hillier* *(69)* performed systematic calculations of this relaxation energy for a large number of gaseous molecules, finding values between 11.7 eV (for CO where the actual *Koopmans* value is 310.7 eV and the experimental $I = 296.2$ eV) and 15.2 eV for carbon, 16.7 to 19.2 eV for nitrogen and 19.3 to 22.3 eV for oxygen. The order of magnitude of the inter-atomic relaxation energy (in addition to the intra-atomic relaxation energy in the isolated atom) is 6 eV but seems to vary somewhat unpredictably from one molecule to another. Because of the charging effect on non-conductors, it is very difficult to determine the additional relaxation energy going from a gaseous molecule to the condensed solid, though it seems to be rather small for organic compounds. One expects rather dramatic decreases for metallic elements compared with the gaseous atoms. However, very few data have been published, with exception of the perhaps somewhat atypical case of $I(4f) = 12.0$ eV in europium atoms (*70*) having the ground configuration $[Xe]4f^7 6s^2$ to be compared with $I^*(4f) = 2.1$ eV in metallic europium (*71*) having the work function 2.6 eV (*72*) and hence $I = 4.7$ eV.

It is possible to incorporate neon, argon, krypton and xenon atoms in the coinage metals copper, silver and gold. *Citrin* and *Hamann* (*73, 74*) found that the $I$ values (relative to *vacuo*) have decreased from 2.2 to 4.0 eV. Interestingly enough, this decrease is the most pronounced for the lighter (and less polarizable) noble gases and is at least 3.5 eV for neon. Nobody can maintain that neon atoms in a metallic lattice assume a large, negative fractional charge, and these results show unequivocally that chemical shifts $dI$ cannot be explained exclusively by a variation of the fractional charge (and the *Madelung* potential) but must contain contributions from inter-atomic relaxation energy.

In fairly ionic potassium(I) compounds, it is improbable that the fractional charge of potassium is much below +1. Table 2 gives the $I'$ values for K$2p_{3/2}$, K$3s$ and K$3p$. In the cases where $\partial$ cannot be determined, the $I$ values are given as higher limits to $I'$. Then, the *Madelung* potential $V_{Mad}$ at the potassium site in the NaCl type halides is given as 50.2 eV$/a_0$ and added to the three $I'$ values. In $K_2PtCl_6$ type cubic crystals, the approximation of $CaF_2$ type 83.6 eV$/a_0$ is applied, and in the perovskite $KNiF_3$ the more doubtful CsCl hypothesis 29.3 eV$/a_0$. This expression is also applied to $KIO_3$ which is approximately a perovskite with $a_0 = 4.46$ Å. In some of the other salts, the *Madelung* potential has been roughly estimated and given as an integer. As pointed out by *Citrin* and *Thomas* (*49*) the difference between the ionization energy of K$3s$ and K$3p$ remains roughly the same as 16.06 eV in gaseous $K^+$. The experimental scattering may not be significant because the most deviating $I'$(K$3s$) (given in parentheses) are perturbed by the adjacent fluorine $2s$ signal. Actually, the distance between the ionization energies of K$2p_{3/2}$ and K$3p$ remains much more constant 276.1 eV with an average deviation 0.2 eV in the 28 examples. It is difficult to verify the corresponding $I(\mathrm{K}2p_{3/2}) = 276.1 + 31.6 = 307.7$ eV in gaseous $K^+$ because such a species only exists in a plasma where it would be difficult to measure $E_{kin}$ of the photo-electrons. There is a weak tendency toward slightly higher values of this difference in the beginning than at the end of Table 2, and it cannot be excluded that $I(\mathrm{K}2p_{3/2})$ is closer to 308 eV in $K^+$.

Anyhow, there is no doubt that interatomic relaxation effects occur in these potassium salts. There are good reasons to believe that the *Madelung* potential is slightly over-estimated by the point-charge model, and whereas the interatomic relaxation may represent only 0.4 eV in KF, it is close to 3 eV in KI, $KIO_3$ and $KIO_4$ and interestingly enough also in $KNiF_3$ whereas the salts of monomeric fluoro complexes have more moderate interatomic relaxation apparently scattered between very small values and perhaps 2.5 eV for $KPF_6$. The very large univalent

Table 2. The ionization energies $I'$ of the potassium $2p_{3/2}$, $3s$ and $3p$ corrected for the charging effect $\partial$, the quantity $C'_{st}$ defined in the text, and $I'$ added to the (in part approximate) Madelung potential of the crystal at the K site

| | $\partial$ | $C'_{st}$ | $I'(K2p_{3/2})$ | $+V_{Mad}$ | $I'(K3s)$ | $+V_{Mad}$ | $I'(K3p)$ | $+V_{Mad}$ |
|---|---|---|---|---|---|---|---|---|
| $KBF_4$ | 2.6 | 2.8 | 299.3 | 306 | 38.8 | 45 | 23.1 | 30 |
| $KPF_6$ | 1.8 | 2.7 | 299.3 | 305 | 37.2 | 43 | 22.9 | 29 |
| $K_2SiF_6$ | 2.6 | 2.4 | 298.8 | 308 | 38.6 | 48 | 22.9 | 32 |
| $K_2BeF_4$ | 3.0 | 2.0 | 298.5 | 308 | 38.6 | 48 | 22.4 | 32 |
| $K_2TiF_6$ | 2.1 | 2.4 | 298.3 | 307 | (34.8) | — | 22.2 | 31 |
| $K_2PtCl_6$ | 1.6 | 3.0 | 298.3 | 306.9 | 38.3 | 46.9 | 22.1 | 30.7 |
| KBr | 1.5 | 3.1 | 298.3 | 305.9 | 37.8 | 45.4 | 21.8 | 29.4 |
| $K_2TaF_7$ | 1.4 | 3.1 | 298.2 | 307 | (35) | — | 22.4 | 31 |
| $KIO_3$ | 2.1 | 3.2 | 298.1 | 304.6 | 38.4 | 44.9 | 21.7 | 28.2 |
| $KIO_4$ | 2.5 | 2.8 | 298.0 | 304 | 37.6 | 44 | 21.8 | 28 |
| KF | 2.5 | 1.7 | 297.9 | 307.3 | 37.8 | 47.2 | 21.7 | 31.1 |
| KI | 3.1 | 1.6 | 297.9 | 305.0 | 37.6 | 44.7 | 21.7 | 28.8 |
| KCl | 2.5 | 2.3 | 297.8 | 305.8 | 37.7 | 45.7 | 21.4 | 29.4 |
| $K_2HAsO_4$ | 1.7 | 2.8 | 297.8 | — | 38 | — | 21.8 | — |
| $KNiF_3$ | 3.2 | 1.3 | 297.7 | 305.0 | (35) | — | 22.0 | 29.3 |
| $KOsO_3N$ | 2.3 | 2.1 | 297.7 | 304 | 37.7 | 44 | 21.7 | 28 |
| $K_2Ru(NO)Cl_5$ | 3.3 | 1.5 | 296.1 | 304 | 36.1 | 44 | 19.9 | 28 |
| $K_2ReBr_6$ | — | <4.8 | <299.3 | <307.3 | <39.4 | <47.4 | <22.9 | <30.9 |
| $KB(C_6H_5)_4$ | — | <5.1 | <298.9 | <302 | <39.0 | <42 | <22.7 | <25 |
| $K_2PtBr_6$ | — | <2.6 | <298.8 | <306.9 | <39.4 | <47.5 | <23.1 | <31.2 |
| $K_2Ni(CN)_4$ | — | <4.9 | <298.8 | — | <38.9 | — | <22.9 | — |
| $K_2ReCl_6$ | — | <3.7 | <298.7 | <307.2 | <39.0 | <47.5 | <22.4 | <30.9 |
| $K_4Fe(CN)_6$ | — | <5.1 | <298.6 | — | <39.0 | — | <22.8 | — |
| $K_2OsCl_6$ | — | <3.8 | <298.4 | <307.0 | <38.7 | <47.3 | <22.7 | <31.3 |
| $K_3Cr(NCS)_6$ | — | <4.8 | <298.4 | — | <38.7 | — | <22.6 | — |
| $KMnO_4$ | — | <5.1 | <298.4 | <305 | <38.7 | <46 | <22.4 | <30 |
| $K_2Pt(SCN)_6$ | — | <4.8 | <298.3 | — | <38.5 | — | <22.5 | — |
| $K_3Fe(CN)_6$ | — | <5.4 | <297.9 | — | <38.1 | — | <21.9 | — |
| $K^+$ gaseous | — | — | 307.7 ? | — | 47.68 | — | 31.62 | — |

anion in $KB(C_6H_5)_4$ is a quite outstanding case with interatomic relaxation close to 6 eV.

Originally, it was argued (*15, 75*) that $I(K2p_{3/2})$ closely follow the order of anions in *Pearson*'s classification (*76, 77*) of hard and soft bases, with the conspicuous exception of iodide. This statement is slightly less clear-cut when correcting for charge effects, but the idea remains that the interatomic relaxation effect is an appropriate measure of *chemical polarizability*, which is not true for the additive molar refractivities (*78, 79*) derived from refractive indices. *Pearson* (*77*) has recently edited a reprint collection about this subject, and the writer (*79*) has reviewed other characteristics of soft central atoms, such as the slightly distorted stereochemistry of copper(II) and palladium(II) complexes and the connections with the hypersensitive pseudoquadrupolar transitions in lanthanides (*80, 81*). It seems (*7*) that the LCAO model with a few orbitals from each atom participating in the chemical bonding is inadequate for describing the complexes of soft central atoms such as Ag(I), Hg(II) and Tl(III) and of ligands such as $PF_3$ bound to central atoms in negative or zero oxidation states, and that one has to take *polarization orbitals* from the continuum into account.

The importance of relaxation energy for chemical shifts in potassium salts makes it clear that $dI$ in transition and post-transition group compounds are strongly influenced by variations of the fractional atomic charge, of the *Madelung* potential, and of the interatomic relaxation. In the *d* groups (*15*) the order of $dI$ follows fairly closely the nephelauxetic series (*33, 34*) of decreasing parameter of interelectronic repulsion with the fluorides at highest $I$ and sulphides and complexes of sulphur-containing ligands with the lowest $I$ for a given oxidation state. However, there is no doubt that $dI$ is determined by partly counter-acting tendencies, and from this point of view is comparable to the spectrochemical series (*13, 21*) of sub-shell energy differences, mainly representing the difference between $\sigma$-anti-bonding and $\pi$-anti-bonding effects on the five *d*-orbitals. Thus, $I$ are unexpectedly high for many iodides and iodo complexes, and also for cyanides. Since this is true (*15*) also for the palladium(IV) complex $Pd(CN)_6^{-2}$ it is hardly due to excessive back-bonding increasing the fractional charge of the central atom. Another, less pronounced discrepancy with the nephelauxetic effect is that chloro complexes such as $PdCl_4^{-2}$ tend to show higher $I$ values than ammonia complexes such as $Pd(NH_3)_4^{+2}$ but this tendency may be connected with smaller $\partial$ in the latter hydrogen-bonded systems. It is interesting to note that nickel(II) oxide constitutes an exception both as far goes a moderately high set of $I$ values as a much less pronounced nephelauxetic effect than in most other oxides.

Most elements (including the *d*-groups) in the oxidation state $z$ have the property (*82*) that $I$ of the gaseous or $I'$of the solid fluoride is approximately $z$ eV higher than of the corresponding oxide. This difference is far smaller in the immediate post-transition elements such as Zn(II), Cd(II) and Hg(II). This fact may conceivably be connected with the usually high coordination number $N = 6{,}8$ and 8 of these elements.

One should not conclude conversely that compounds are exclusively electrovalent because the $I$ values correspond to the *Madelung* potential for full ionicity given in Table 1. However, it is interesting to note (*2*) that such a description is not perfect for beryllium(II) compounds having $I'(\mathrm{Be}\,1s)$ varying between 120.8 eV for $BeF_2$ and 118.5 eV for the carbonate complex $[Co(NH_3)_6]_2[OBe_4(O_2CO)_6]$ to be compared with 153.89 eV for $Be^{+2}$, whereas $V_{\mathrm{Mad}} = 23.9$ eV is no bad description of MgO having $I'(\mathrm{Mg}\,2p) = 54.1$ eV and $Mg^{+2}$ has $I = 80.14$ eV. $I'(\mathrm{O}\,2p)$ close to 10 eV corresponds to a negative $I$ of $O^{-2}$ after correction for the *Madelung* potential, as well-known from the *Born-Haber* cycle applied to oxides. Generally, the variation of the full *Madelung* potential over-estimates the differences between different salts formed by a definite cation.

## E. Comparison with X-ray Spectra

The photon energies of X-ray lines emitted by closed-shell systems are generally identified with the difference between two ionization energies $I_0$ and $I_f$. This approach was used by *Cauchois* (*83, 84*) for compiling inner-shell energies said to be accurate within 5 or 10 eV, a modest percentage of the photon energies measured. *Bearden* and *Burr* (*85*) performed a statistical analysis of the emission line data available and proposed $I$ values with precisions given, frequently below 1 eV. This precision runs into various problems, such as the *Fermi* levels of metallic anti-cathodes; ideally, the tables are supposed to treat the elements, but certain compromissions regarding oxides etc. were made. A quite fundamental question is that X-ray emission obeys the *Franck-Condon* principle, conceivably destroying the linear subtraction $(I_0 - I_f)$ to the extent of 1 or 2 eV. Only recently, vibrational structure was reported (*86, 87*) for a gaseous molecule, but this does not prevent unresolved vibrational broadening from occurring in solids. When the total spin quantum number $S$ of the groundstate is positive because of the presence of a partly filled $d$ or $f$ shell with sufficiently small average radius, the electron configurations corresponding to only $(4l+1)$ electrons in one of the inner shells get separated in many energy levels which may be distributed over 10 to 20 eV. Frequently, this separation goes undetected because of the selection rules for electric dipole transitions and the numerical questions of oscillator strength only allow a few of the conceivable transitions to give prominent emission lines.

A practical weakness of determining $I$ from X-ray emission spectra is that the inner shells are the best studied, and the uncertainties of combining various transitions tend to pile up on the penultimate shells. A typical case is $I'(\mathrm{Th}\,6p_{3/2})$ found (*15*) at 23.6 eV in $ThF_4$ and at 20.7 eV in $ThO_2$ where the Uppsala tables (*12*) give $I^* = 43$ eV. However, *Fuggle et al.* (*88*) finally located this level at $I^* = 16.6$ eV in metallic thorium and at the marginally higher $I^* = 16.8$ eV in metallic uranium. The reason behind the large uncertainties (not only percentage-wise but even in eV) of penultimate orbitals is that X-ray spectrography normally is performed between 0.1 Å (the 1*s* ionization limit of uranium) and 5 Å or 2470 eV (close to the 1*s* limit of sulphur) because the experimental difficulties working with the soft X-rays above 5 Å are very large (*89*). Since it is also difficult to work in the very far ultra-violet below 120 Å (100 eV) the intermediate region (which is now opening up to absorption studies using the continuous spectrum (*90—93*) emitted by a synchrotron) was brillantly replaced by photo-electron spectrometry of solids.

There is sporadic evidence that professional X-ray spectroscopists tend to consider the photo-electron people as opportunistic arrivists, preferring a "quickie' to a long romance, but it is also true that the mutual interaction of these two subjects has revived X-ray spectrometry which showed signs a few years ago of beginning fossiliation, like macroscopic crystallography and study of the Asteroïds. Anyhow, a most interesting meeting was held July 1974 in Otaniemi close to Helsinki, where the X-rays from atoms (having collided with a very rapid ion) lacking from one to six $2p$ electrons (*94*), the molecular orbitals formed from the $1s$ orbitals by the collision of 0.5 MeV $Ne^+$ with Na (*95*) and the hypersatellites due to the simultaneous absence of both $1s$ electrons (*96*) were discussed, together with many other exotic subjects.

Chemical shifts on X-ray emission lines are an order of magnitude (*97*) smaller than $dI$, essentially because the central field $U(r)$ is changed by a constant seen from the point of view of the two inner shells, like it would be the case for a completely external *Madelung* potential, but nevertheless, *Stelling* (*98*) in Lund detected chemical shifts in both emission and absorption spectra of compounds of sulphur and chlorine changing their oxidation states by eight units. These observations included the two non-equivalent sulphur atoms in thiosulphate $S_2O_3^{-2}$ which later became the first example (*99*) found in Uppsala of highly different $I$ of the same element. It is very characteristic (*15*) that the distance 6.0 eV between the two $I(S2p)$ of $SSO_3^{-2}$ is decreased to 5.4 eV in the gold(I) thiosulphate complex $O_3SSAuSSO_3^{-3}$ to a smaller extent than to 4.3 eV in the tetrathionate $O_3SSSSO_3^{-2}$ containing two sulphur atoms of each kind.

The trouble with X-ray absorption spectra (frequently showing chemical shifts up to 10 eV) is that the bands are broad and asymmetric (edges) producing methodological difficulties when photographed and, even more significantly, that the electron from the inner shell (which is ejected with a definite $E_{kin}$ in a photo-electron experiment) is excited with an appreciable oscillator strength to an acceptor orbital with opposite parity of the inner shell. In some instances, the strongest transitions go to recognizable, empty or partly filled shells such as (neglecting closed shells) $3d^{10}4f^q \rightarrow 3d^9 4f^{q+1}$ (*100, 101*) and $4d^{10}4f^q \rightarrow 4d^9 4f^{q+1}$ (*102*) in the metallic lanthanides (including lanthanum) and their compounds. $3d$ group complexes show transitions from $1s$ to empty orbitals having the same symmetry types in the local point-group as the three $4p$ orbitals. However, in actual practice, complicated organic ligands produce additional perceptible acceptor orbitals (*103, 104*).

In all of these excitation processes, one meets the same problem as in electron transfer spectra (*4, 105*) that the difference of ionization energy of the stronger bound orbital $b$ and the less strongly bound $a$ is corrected by the interelectronic repulsion $A_*(a,a)$ in the $a$ orbital and by the *charge-separation effect* (*4*) represented by the hole-electron attraction $A_*(a,b)$:

$$h\nu = I(b) - I(a) + A_*(a,a) - A_*(a,b) \ . \tag{20}$$

As the matter of fact, $I(a)$ of empty orbitals can be defined as the sum of the electron affinity and $A_*(a,a)$. This problem also occurs in X-ray emission involving the loosest bound electrons jumping down in an inner vacancy. As we already discussed after Eq. (7) the last term of Eq. (20) is essentially $\langle r^{-1} \rangle$ of the outer electron being attracted as if (*23*) the element had the subsequent atomic number $(Z+1)$. In metals, the bottom of the partly occupied conduction band tails off toward more negative energies than in the groundstate without inner-shell vacancies (this problem seen from the point of view of the relaxation of the surrounding electron density by photo-ionization of a metal is discussed by *Watson* and *Perlman* in this volume) and in compounds, transitions from filled penultimate orbitals to an inner shell can be seen in emission ("crossed transitions" involving different atoms, like the electron transfer spectra in the visible and the ultra-violet) on the condition that the MO written as a LCAO eigen-vector contains appreciable squared amplitudes of a shell having allowed electric dipole transitions within the same atom to the vacancy. Thus, $MnO_4^-$ shows X-ray emission (*106*) from several MO containing Mn4*p* contributions to a Mn1*s* vacancy, and $CaF_2$-type $ThO_2$ from an orbital (*107*) having sufficient 5*f* (*xyz*) character to show sharp transitions to a 3*d* vacancy. It must be remembered that the major part of these MO is formed by O2*p* from the adjacent 4 or 8 oxygen atoms.

However, the X-ray absorption spectra of non-*f*-group compounds show another sophistication quite unaccessible to photo-electron spectra induced by high-energy photons. The electron from the inner *nl*-shell absorbing the X-ray arrives in the continuum (and not in a partly filled 4*f* shell) and produces its own favoured combination of the infinite number of continuum states, preferentially with the quantum number $(l+1)$. This selection by a electric dipole transition is rather similar to the determination of polarization orbitals (*7, 78*) in the deformation of the groundstate wave-function in the presence of a (weak laboratory) electric field, or to the *d*-orbitals needed to explain the inversion barrier and non-planarity of the ammonia molecule (*108*) or, to a certain extent, the continuum orbitals producing the main part of the correlation energy by two-electron substitutions in the electron configuration (*4*). It is quite characteristic in all of these cases that the continuum orbitals have average radii comparable or only slightly larger than the orbital of the ground configuration to be modified. Such orbitals are not at all comparable with the highly excited *Rydberg* orbitals with high $n$ having average radii approximately $n^2$ bohr, and their energy is to large extent angular kinetic, according to Eq. (12) $l(l+1)\langle r^{-2}\rangle/2$ atomic units. This explains the phenomenon observed by *Cauchois* (*109*) that the 3*d* absorption edge of metallic thorium is situated roughly at the *Fermi* level where empty 5*f* orbitals are available to accept the electron with high probability, whereas the main part of the 3*d* absorption in metallic gold occurs about 20 eV above the *Fermi* level. Hence, very few conduction electrons in gold have $l=3$, and they are rather 6*s* with a tiny amount of 5*d* character.

However, this selection of continuum orbitals with low average radii and approximately well-defined $l$ also occurs in gaseous molecules, and was first found in octahedral $SF_6$ (*110*). Here, the S2*p* absorption goes to a totally symmetric $a_{1g}$ orbital, to *d*-like $t_{2g}$ at 5 eV and $e_g$ orbitals at 15 eV higher energy, as found by *Vedrinski* (*111*) using the "Multiple scattering (X$\alpha$) theory" of *Slater* and *Johnson* (*112, 113*). *Dehmer* (*114*) discussed the general *barrier effects* connected with the fact pointed out by *Maria Goeppert-Mayer* (*115*) that $\mathscr{P}_{\text{ang}}$ of Eq. (12) added to the central field $U(r)$ frequently has two minima, an outer shallow minimum situated roughly at $l(l+1)$ bohr$/(z+1)$ where $z$ is the ionic charge, and an inner, deep minimum accommodating the 4*f* electrons in the lanthanides. Most of the oscillator strength of the X-ray absorption is concentrated on transitions to orbitals (discrete or continuum) localized in the inner minimum. Thus, gaseous and solid xenon show $4d \rightarrow f$ transitions where the *f* orbital has no similarity with the widely extended 4*f* orbitals known from configurations such as $5p^5 4f$ but rather is a continuum orbital with $l=3$ localized within the concentric barrier.

The same is true for iodides and the isoelectronic caesium(I). It is also recognized now (*116*) that crystalline iodides have $5p \rightarrow 6s$ and $5p \rightarrow 5d$ inter-shell transitions like xenon condensed together with other noble gases or with various molecules.

These questions are connected with the relative intensities of photo-electron signals (compared with fluorine 1*s* (*117*)) induced by 1486.6 (*118*) or 1253.6 eV photons (*119*). Since the *ns* or *np* absorption edge according to *Kramers* has a molar extinction coefficient $\varepsilon$ or cross-section (above the critical $I = h\nu$) proportional to $\nu^{-3}$ one would expect (*118*) the intensities to be $(2j+1)\ I^2$ times a constant, where $(2j+1)$ is the number of electrons to produce the signal. However, *d* shells give stronger, and *f* shells far stronger signals than predicted by the *Kramers* relation. This is understood (*120, 121*) as a by-product of the angular kinetic energy shifting the absorption maxima to some hundred eV above the edge. When such maxima are called "delayed" (*114*) it is not with respect to time, but on the $h\nu$ scale. Thus, the X-ray absorption of 4*f* electrons go mainly into continuum orbitals with $l = 4$. It turns out (*118*) that the signal intensities are roughly proportional to $(2j+1)\langle r^{-2}\rangle$. We return to this question in Table 4.

## II. Multiplet Structure and Satellites

### A. Multiplets in Intermediate Coupling

The first example of structure due to differing interelectronic repulsion in levels having $(S+{}^1/_2)$ and $(S-{}^1/_2)$ formed by ionization of a system with positive $S$ in the groundstate was detected in Uppsala (*12, 58*) in the oxygen molecule $O_2$ showing two $I$(O 1*s*) signals at 543.1 and 544.2 eV, the former twice as strong as the latter. The two ionized situations are quartet $(S={}^3/_2)$ and doublet $(S={}^1/_2)$, respectively. It was one of the early triumphs of MO theory that *Lennard-Jones* demonstrated that $O_2$ has no conventional double bond like ethylene, but cannot avoid having a triplet $(S=1)$ groundstate with two of the four possible $\pi_g$ electrons in the lowest MO configuration. The two other singlet levels $(S=0)$ of this configuration correspond to terrestrian *Fraunhofer* lines in the infra-red and the red, and the weak, simultaneous transitions on two $O_2$ molecules (*122*) produce the narrow absorption bands and sky-blue colour of liquid oxygen. It was not obvious that the $\pi_g$ orbitals carrying the uncompensated spin (and constructed from 2*p* orbitals in the LCAO approximation) would show detectable effects of interelectronic repulsion with a 1*s* vacancy. The separation between the two levels of the ionized system is $2K(1s, \pi_g)$ neglecting coupling with the two singlet levels of the ground configuration. Since $K(a,b)$ is the electrostatic interaction between the product of two one-electron wave-functions $\psi_a \psi_b$ and itself, the experimental value $K(1s,\pi_g) = 0.55$ eV indicates a certain co-existence of the two orbitals at the same points of our three-dimensional space. An alternative explanation of co-exciting one of the two singlets (at 0.98 and 1.64 eV) of $O_2$ is not plausible, because the same type of separation is found in the N 1*s* signals of NO (*12, 58*) having no excited level at 1.5 eV. Seen from the point of view of MO theory, the whole problem is rather much in a mess, because the presence of a 1*s* vacancy on one of the two oxygen atoms is not symmetry-adapted. However, the overlap integral $\sim 10^{-5}$ between the two 1s orbitals is so small (*123, 124*) that there is no doubt that the migration time would be hundreds times longer than the primary process of photo-ionization. It is possible to get an impression of the relative delocalization of the orbitals carrying the uncompensated spin density by noting that the two molecules NO (*58*) and *t*-butylnitroxide ($((CH_3)_3C)_2NO$ (*125*) both having $S = 1/2$ have much larger separations of the N 1s than of the O 1s signal. Whereas all molecules containing an odd number of electrons have $S$ at least $1/2$, higher $S$ values are rather rare outside the transition groups though calculations (*126*) on $C_2^+$ indicate a groundstate with $S = 3/2$ belonging to a MO configuration with two $\pi_u$ and one $\sigma_g$ orbital half-filled.

The situation is entirely different in the 3*d* group compounds frequently (*21*) having $S$ of the groundstate up to $5/2$, and in the 4*f* group up to $7/2$. Actually, the barycentre of all states belonging to a definite configuration containing one partly filled shell $l^q$ and having a given value of $S$ contains a contribution to the interelectronic repulsion

$$\frac{q(q-1)}{2} A_*(l,l) + \left[\frac{3q(4l+2-q)}{16l+4} - S(S+1)\right] D \qquad (21)$$

where the *spin-pairing energy parameter* $D$ is a definite linear combination (*4*) of the $l$ different *Slater-Condon-Shortley* (*25*) parameters of interelectronic repulsion $F^2, F^4, \ldots, F^{2l}$ or, equivalently, of *Racah* (*127*) parameters such as $^9/_8 E^1$ for $l = 3$. $D$ turns out (*21*) to be $(2l+3)K_{av}(l,l)/(2l+2)$ where the parameter $K_{av}(l,l)$ is the average value of the $(4l^2+2l)$ different $K$ parameters occurring for the closed shell with $q = 4l+2$. The order of magnitude of $D$ is 0.8 eV in the 3*d* and 4*f* groups and about half as large in the 4*d*, 5*d* and 5*f* groups (*21*). Correspondingly, half-filled ($q = 2l+1$) or nearly half-filled shells are stabilized according to Eq. (21) to a considerable extent ($-20D/3$ for $d^5$ and $-168D/13$ for $f^7$) on the condition of having the maximum value of $S$. The barycentre of all states having $(S-1)$ is situated $2DS$ above the barycentre of $S$.

The simplest inner shell to ionize is a *ns* orbital. *Van Vleck* (*7, 25*) demonstrated that $l^q$ with positive $S$ has the levels of $sl^q$ having $(S + 1/2)$ situated $(2S+1)$ $K(s,l)$ below the levels having $(S - 1/2)$. The 1s signals have not been studied in àny *d*-group compound with positive $S$ because of their high energy. It is possible to study 2s and 3s signals of the 3*d* group, though the signals are rather large (the *np* electrons rapidly perform *Coster-Kronig* transitions to the *ns* vacancy) and comparatively weak (*118, 119*). Nevertheless, two adjacent signals can be seen in the $3d^5$ systems Mn(II) and Fe(III) (*70, 128, 129*), the $3d^3$ systems Cr(III) (*128–130*) and Mn(IV) (*128*) and in $3d^7$ Co(II) and $3d^8$ Ni(II) (*15*). In all of these cases, $K(3s,3d)$ and $K(2s,3d)$ scatter between 0.8 and 1.2 eV which, surprisingly enough, is higher than $K_{av}(3d,3d) = (6D/7)$ from Eq. (21). The *Hartree-Fock* functions of *Watson* (*4, 131*) give $K(2s,3d) = 0.8$ eV for $Cr^{+3}$ and 1.0 eV for $Ni^{+2}$ and $K(3s,3d) = 2.2$ eV and 2.6 eV in the two cases. We return below in Chapter II E to the specific problem of three $I$(Ni 3s) signals reported in $3d^8(S=1)$ nickel(II) compounds. *Cohen, Wertheim, Rosencwaig* and *Guggenheim* (*132*) studied $I$(M 4s) and $I$(M 5s) of lanthanide $MF_3$ and found $K(4s,4f) = 1.1$ eV and $K(5s,4f)$ between 0.4 and 0.5 eV.

The situation is far more complicated in the case of ionization of an inner $np$ shell. The overall spreading of $I(\mathrm{M}3p)$ in the $3d$ group has the same order of magnitude (*15, 128, 129*) as $I(\mathrm{M}3s)$ but one sees shoulders and frequent indications of more than two signals. One cannot immediately use *Van Vleck's* formula for $p$ ionization, since configurations such as $3p^5 3d^q$ have a complicated distribution of energy levels. Because a requirement for strong signals is high fractional parentage of the groundstate of $3d^q$ in the ionized level, it can be argued that the $^6S$ level of $3d^5$ in manganese(II) and iron(III) compounds preferentially should ionize to $^7P$ and $^5P$ of $3p^5 3d^5$ (neglecting closed shells). It turns out that several $^5P$ terms (*129*) also having contributions to the fractional parentage from $3d^5$ terms such as $^4P$ and $^4D$ (able to form $^5P$ by vector-coupling with $^2P$) show up in the photo-electron spectrum. An additional complication in $I(\mathrm{M}2p)$ is the strong spin-orbit coupling separating $2p_{1/2}$ and $2p_{3/2}$ to an extent smoothly increasing from 9.2 eV in chromium to 19.9 eV in copper. *Bonnelle* and the writer (*133*) showed that the $J$-level distribution of the configuration $2p^5 3d^9$ obtained from X-ray absorption spectra of thin films of NiO or as the initial state of X-ray emission (by fluorescence) of CuO can be explained by the equations (*25*) for intermediate coupling using parameters of interelectronic repulsion derived from *Watson*'s *Hartree-Fock* functions (*131*). However, we wait discussing the structure of the general case $2p^5 3d^q$ to chapter 2E.

The $I(\mathrm{M}4p)$ of lanthanides have not been much studied because the signals are so weak. Curiously enough, $I(\mathrm{M}5p)$ does not seem to broaden to a significant extent (excepting an increasing separation of $5p_{1/2}$ and $5p_{3/2}$) going from La(III) over Gd(III) to Lu(III) though this may be due to severe discriminations by coefficients of fractional parentage.

On the other hand, the $I(\mathrm{M}4d)$ show a very complicated behaviour. The configuration $4d^9 4f^q$ is also the excited states of X-ray absorption spectra of metallic lanthanides (*102*) or M(III) compounds with groundstates belonging to $4f^{q-1}$ It is known to contain 10 times as many orthogonal states as $f^q$ having in spherical symmetry

| | | | | | | | | | |
|---|---|---|---|---|---|---|---|---|---|
| $q =$ | 0,14 | 1,13 | 2,12 | 3,11 | 4,10 | 5,9 | 6,8 | 7 | |
| states | 1 | 14 | 91 | 364 | 1001 | 2002 | 3003 | 3432 | (22) |
| $J$-levels | 1 | 2 | 13 | 41 | 107 | 198 | 295 | 369 | |
| $S,L$-terms | 1 | 1 | 7 | 17 | 47 | 73 | 119 | 119 | |

It certainly represents a case of intermediate coupling, because the spin-orbit separation between $4d_{3/2}$ and $4d_{5/2}$ is 10.0 eV in $4f^{14}$ Lu(III) and 10.7 eV in Hf(IV) to be compared with photo-electron signals in the $4d$ region of ytterbium(III) compounds (*2, 15*) distributed over an interval above 20 eV. *Signorelli* and *Hayes* (*134*) calculated the positions of the 20 $J$-levels of $4d^9 4f^{13}$ in intermediate coupling and found them distributed fairly evenly over an interval of 27.4 eV. There is no clear-cut separation between the 12 levels due to ionization of $4d_{5/2}$ and 8 higher levels due to $4d_{3/2}$. It is interesting to compare with the coupling with outer electrons, where *Bryant* (*135*) find the 10 levels of $Yb^{+2}$ belonging to the configuration $4f^{13}5d$ having $^2F_{7/2}$ parentage between 4.14 and 5.40 eV above the groundstate $4f^{14}$ whereas the 10 other levels of $4f^{13}5d$ with $^2F_{5/2}$ parentage are located between 5.73 and 6.66 eV. The barycentres of the two sets of $J$-levels are separated 1.25 eV like all other instances (*136*) of $4f^{13}$ defining (*21*) the conditional oxidation state Yb[III] though the separation is not so conspicuous as the four $J$-levels of $4f^{13}6s$ of $Yb^{+2}$ (*135*) situated at 4.29, 4.33, 5.56 and 5.60 eV.

The $4d$ regions of most of the lanthanides (*15, 137*) show broad hills obviously due to adjacent, superposed signals. We run into the problem first pointed out by *Fadley* and *Shirley* (*70*) in the case of $LuF_3$ that the half-life of the ionized system containing a $4d$ vacancy is so short, due to *Coster-Kronig* transitions $4d^9 4f^{14} \rightarrow 4d^{10}4f^{13}$ that the intrinsic half-width $\delta$ according to the *Heisenberg* uncertainty principle is 1.9 eV. This value (*15*) remains almost constant between 1.9 and 2.3 eV in the elements between lutetium and uranium. This problem of a large density of adjacent $4d$ signals in the middle of the lanthanides cannot be solved by an ameliorated instrumental resolution (as the case of $4f^q \rightarrow 4f^{q-1}$ ionization discussed in Section III and in another volume (*138*) of Structure and Bonding). In this volume, *Cox* treats the calculation of coefficients of fractional parentage.

It may be worthwhile to mention (*15*) that gadolinium(III) compounds have two $4d$ signals separated to the extent of 5.3 eV whereas the $3d_{3/2}$ and $3d_{5/2}$ signals are only weakly asymmetric. An interpolation between the spin-orbit coupling in $4d$ being 2.6 eV in Ba(II) and 10.0 eV in Lu(III) may suggest that the two prominent (but broad) maxima in Gd(III) are the two sets of six and four $J$-levels obtained by coupling $^8S_{7/2}$ (the highly isolated groundstate of $4f^7$ )with one $4d_{5/2}$ or $4d_{3/2}$ vacancy. The alternative description of $^9D$ and $^7D$ terms feasible in the lowest configuration $4f^7 5d$ of $Gd^{+2}$ (*139*) only differs superficially by the transfer of one level with $J=1$ from the lower to the higher manifold. Nevertheless, the calculation (*134*) for Yb(III) shows that inter-shell effects of interelectronic repulsion must be very important in the general $4d^9 4f^q$ configuration with small, and comparable, average radii of the $4d$ and $4f$ shells. It may be noted that $\zeta_{nl}$ to a fair approximation is (1 rydberg/$137^2$) times $Z_{\text{eff}}\langle r^{-3}\rangle$ but that it is not generally meaningful to evaluate $\langle r^{-3}\rangle$ from the hydrogenic expression Eq. (13) for inner shells.

## B. Electron Transfer Satellites in Lanthanum(III)

One would not expect multiple photo-electron signals to occur by the ionization of an inner shell in a diamagnetic closed-shell compound. A rather unique case is the $3d_{3/2}$ and $3d_{5/2}$ signals (at 16.9 eV distance) of the oxygen-containing lanthanum(III) compounds $La(IO_3)_3$, $La_2(SO_4)_3$, $La_2(CO_3)_3$, $LaAsO_4$, LaOCl, $LaCrO_3$ and $La_2O_3$ accompanied (*140*) at 3.1 to 3.7 eV higher $I$ by a satellite of comparable intensity. In $LaF_3$ which was also studied by *Wertheim et al.* (*141*) the satellites have less intensity and correspond to shoulders at 4.4 eV higher $I$. In the xenon atom (*58*) $I(\text{Xe}3d_{3/2}) = 689.0$ and $I(\text{Xe}3d_{5/2}) = 676.4$ eV. In the isoelectronic caesium(I) salts, two narrow $3d$ signals are separated (*15*) by 14.05 eV and in barium(II) by 15.3 eV without any sign of four signals. The satellites are not *Auger* signals because the spectra are identical with 1486.6 and 1253.6 eV photons.

*Berthou* and the writer (*140*) ascribe the satellites to an inter-atomic shake-up process, where a $2p$ electron from adjacent oxygen or fluorine atoms is transferred to the empty $4f$ shell of lanthanum at the same time as the $3d$ electron is ejected. Hence, the ionized system contains three partly filled shells, $2p^5$ of a ligating atom and $3d^9 4f$ of the lanthanum. The authors from *Bell* (*141*) suggested transfer of an electron from the valence band of $LaF_3$ to the empty $4f$ band, which is essentially the same explanation, though the energy band model (*4*) is in serious difficulties in metallic lanthanides and their compounds because of the huge difference (*142*) above 7 eV between the ionization energy and the electron affinity of the partly filled $4f$ shell.

La(III) lacking an electron in an inner shell has a central field $U(r)$ rather similar to cerium(IV). It is characteristic (*105, 143*) for the electron transfer spectra of the yellow and red Ce(IV) complexes that they have much higher intensities than of trivalent lanthanides in the ultra-violet such as Sm(III), Eu(III), Tm(III) and Yb(III) complexes of reducing ligands. These higher transition probabilities correspond to larger overlap (or more precisely, greater spatial co-existence) of the empty $4f$ shell of Ce(IV) with filled ligand orbitals than is the case in the later lanthanides. This hypothesis is verified by *Signorelli* and *Hayes* (*134*) finding stronger satellites at shorter distance in anhydrous $LaCl_3$ and with the more reducing ligands in $LaBr_3$ the satellites almost coincide with the main signals, which are now devaluated to shoulders at lower $I$. Recently, *Suzuki*, *Ishii* and *Sagawa* (*144*) isolated the $4d$ satellites by deconvolution of the photo-electron spectrum, at 3.4 eV higher energy in $LaF_3$, 3.1 in $LaCl_3$ and 2.9 eV in $LaBr_3$ with decreasing distance and higher intensity in direction of the reducing bromide. $3d$ satellites have been reported (*145*) in metallic lanthanum.

The general consensus is that cerium(III) also shows electron transfer satellites in the $3d$ (*134, 145*) and $4d$ (*144*) regions. However, this $4f^1$ system has $S = {}^1/_2$ and it is not absolutely excluded that the twin $3d$ signals (*15*) separated by some 3 eV are due to interelectronic repulsion differences. A rather irritating problem is that no spectra of nominal cerium(IV) compounds such as $CeO_2$ and $CeF_4$ have been reported without a set of apparent Ce(III) signals, either due to accidental coincidences or to partial superficial reduction (perhaps under the influence of $CO_2$ in the air) to cerium(III). One expects the electron transfer satellites gradually to disappear in the following lanthanides because of decreased average radii of the $4f$ shell. Actually, the $3d$ signals (*15, 28*) of praseodymium(III) and neodymium(III) compounds have the highly unusual property of showing shoulders at *lower* $I$, which may either be a feature of the $J$-level distribution of $3d^9 4f^2$ and $3d^9 4f^3$, respectively, or as suggested by *Christiane Bonnelle*, electron transfer satellites being so energetically favoured as to have lower $I$ than the main signal.

The closed-shell system thorium(IV) shows no electron transfer satellites (*15*). Protactinium is too rare and too radioactive to be measured by us, but certain uranyl compounds such as $RbUO_2(O_2NO)_3$ show satellites about 3.5 eV above $I(U 4f_{5/2})$ and $I(U 4f_{7/2})$ probably due to electron transfer to the empty $5f$ shell. The somewhat peculiar photo-electron spectra of uranium compounds are further discussed Chapter II F.

### C. Satellites in Copper(II) and in Carbonyls and Cyclopentadienides

*Novakov* (*146*) found the $Cu2p_{1/2}$ and $Cu2p_{3/2}$ photo-electron signals of many copper compounds accompanied by satellites at 8 to 10 eV higher $I$. It is now firmly established (*15, 147*) that closed-shell copper(I) compounds do not show such satellites, but only Cu(II). It would be difficult to ascribe so large an energy difference to interelectronic repulsion which is known (*133*) to separate the two sets of $2p^5 3d^9$ to the extent of 2 to 3 eV. This assignment can better be applied to a shoulder (*15*) seen in several Cu(II) compounds with $I$ some 2 eV above the main $Cu2p_{3/2}$ signal. The satellites at higher $I$ is ascribed by some authors (*146–150*) to shake-up of the type $3d \rightarrow 4s$ or $3d \rightarrow 4p$ (simultaneous ejection of two electrons corresponding to a shake-off process would need more energy according to Eq. (17) and the observed $I'$(Cu3d) close to 11 eV) whereas *Kim* (*151, 152*) and *Carlson et al.* (*153*) believe in electron transfer to the vacant position in the $3d(x^2 - y^2)$ orbital also producing the $3d^9$ transitions below 2.5 eV and the electron transfer bands of copper(II) complexes (*13, 14*).

Besides the more general question why many other oxidizing central atoms containing partly filled d shells and showing electron transfer bands in the near ultra-violet (*105*) below 5 eV, such as Co(III), Rh(III), Pd(II) and Pt(IV) do not show satellites at their photo-electron signals, it is a problem for the electron transfer assignment of the Cu(II) satellites that their distance (*147*) above the main $Cu2p_{3/2}$ signal *increases* from 7.5 eV in $CuF_2$ to higher values such as 8.7 eV CuO and in more covalent compounds. It may be noted that several Cu(II) complexes seem to give two or three adjacent satellites, though it is difficult to obtain a satisfactory counting statistics of the electrons with the high back-ground. It is important (*151, 154*) that not only $Cu2p$ but also Cu3s, $Cu3p$ and $Cu3d$ show satellites about half as strong as the main signals.

The writer tends toward ascribing the copper(II) satellites to some kind of shake-up process enhanced by the specific characteristic of the loosest bound $3d$-like electron (with the angular function proportional to $(x^2 - y^2)$) being strongly anti-bonding and delocalized to a large extent on the ligands, with concomitant large $\langle r \rangle$. The electron transfer satellites obeying monopole selection rules (*i.e.* the electron conserves its symmetry type in the point-group, and specifically $l$ in spherical symmetry) according to *Kim* (*152*) were expected to show up in the complexes of the other oxidizing central atoms. It is by no means firmly established that shake-up processes always are monopole transitions, as also said by *Hamnett* (*155*) in a discussion remark. Thus, the $nd(3z^2 - r^2) \rightarrow (n+1)s$ is an allowed monopole transition between two totally symmetric ($a_{1g}$) orbitals in the point-group $D_{4h}$ characterizing the quadratic Ni(II), Pd(II) and Pt(II) complexes not showing perceptible satellites.

Some metallo-organic compounds show satellites. This is for instance the case (*156–158*) for the neutral carbonyls $Cr(CO)_6$, $Fe(CO)_5$, $Ni(CO)_4$, $Mo(CO)_6$ and $W(CO)_6$. However, it is the C1*s* and O1*s* signals of the ligand showing satellites at 5 to 6 eV higher $I$, whereas the inner shells of the central atom does not exhibit satellite signals. These carbonyls have very intense transitions in the ultra-violet close to 4 eV, and it might be argued that the satellites are indirect optical excitations produced by a shake-up process promoting one of the d-like electrons of the central atom (having $I$ somewhere (*6*) between 8.5 and 8.8 eV) to an empty $\pi^*$ orbital. This is almost the same explanation as electron transfer satellites where an 1$s$ vacancy on carbon or oxygen attracts a $d$-like electron. However, it is clear from our discussion that the satellites have distances frequently rather different from the optical excitation energies, and in particular, the modified central field of La(IV) lacking an internal electron corresponds to a much smaller energy difference than the electron transfer bands of La(III).

*Hamnett* (*155*) pointed out that cyclopentadienides $M(C_5H_5)_2$ show satellites of the M2$p$ signals for those M = Mn and Ni containing two anti-bonding 3$d$-like electrons, but M = Fe without anti-bonding electrons does not show satellites. This observation may constitute a valuable analogy to the specific Cu(II) satellites. It may be noted that $Cr(C_6H_6)_2$ does not (*158*) show C1$s$ satellites in contrast to the other $3d^6$ system $Cr(CO)_6$.

### D. Scandium(III) and Titanium(IV) Oxides and Fluorides

One of the difficulties for discussing significantly the satellites is that they seem to have rather different origins. A new class was reported by *Wallbank, Johnson* and *Main* (*159*) finding satellites at 11 to 13 eV higher $I$ to the M2$s$, M2$p_{1/2}$ and M2$p_{3/2}$ signals of scandium(III) and titanium(IV) oxides and fluorides. The most precisely determined distances are 12.2 eV for $ScF_3$, 11.3 eV for $Sc_2O_3$ and 13.3 eV for $TiO_2$ in the case of M2$p_{3/2}$. These authors suggest electron transfer shake-up from neighbour atom 2$p$ to the empty 3$d$ shell of M. This interpretation is compatible with the weakly oxidizing character of Sc(III), but it is difficult to understand why the satellite distance does not decrease in the more oxidizing Ti(IV), and even more why such satellites are not detected (*15, 130*) in V(V), Cr(VI) and Mn(VII) though it must be admitted that permanganates are readily decomposed to Mn(IV) giving two sets of signals, though this impurity was the least pronounced in $CsMnO_4$. Titanium(IV) has been further studied (*272, 275*).

The satellite distance in Sc(III) and Ti(IV) is almost sufficient for shake-off of a ligand 2$p$ electron without interference of Eq. (17). The writer's diagnosis is $2p \rightarrow 3s$ shake up intensified by the covalent bonding of the 2$p$ orbitals with the empty M3$d$ orbitals since the local $\langle r^{-2} \rangle$ is so important for the signal intensity (*118*). Chlorides should show comparable $3p \rightarrow 4s$ shake-up at 10 eV and bromides $4p \rightarrow 5s$ at 9 eV. This seems indeed to be the case (*153*) and this class of satellites is related the satellites to the 4$s$ signal (producing $[Ar]3d^{10}4s4p^6$) of rubidium(I) salts, which were ascribed (*160*) to higher excited configurations of Rb(II) such as $[Ar]3d^{10}4s^24p^45s$ corresponding to the simultaneous shake-up $4p \rightarrow 5s$ and *Auger* process $4p \rightarrow 4s$. *Spears, Fischbeck* and *Carlson* (*161*) propose the alternative $[Ar]3d^{10}4s^24p^44d$ for the signal 26.0 eV above $I$(Rb4$p$) in RbCl. These authors find satellites of the Kr3$p$ and Kr3$d$ signals of krypton due to the configuration terminating $4s^24p^44d$ at 20.4 eV higher energy, and in xenon the 3$d$ and 4$d$ signals accompanied at 16.9 eV higher $I$ due to configurations terminating $5s^25p^45d$.

Metals are characterized by a plasmon frequency $E_P$ proportional to the square-root of the number of conduction electrons per Å$^3$. Metallic lithium (*162*) shows $E_P = 7.5$ and $2E_P = 14.8$ eV above $I^*$(Li1$s$) = 54.8 eV and sodium (*162, 163*) five almost equidistant plasmon signals with consecutive distances 5.8 to 5.9 eV above $I^*$(Na 1$s$) = 1071.7 eV. Aluminium (*162, 164*) has also five plasmon signals with $E_P$ close to 15.5 eV above I*(Al 2$s$) = 118.4 eV. It is discussed (*162*) why silver has $E_P = 3.9$ eV, half the theoretical value. Metallic cadmium, indium and tin have $E_P$ slightly below the theoretical values 11.3, 12.5 and 14.3 eV. It is noted that these plasmon signals only are seen under painstaking experimental conditions of ultra-high vacuo.

The plasmon signals represent a border-line case between shake-up and inelastic scattering. In non-conductors, one observes comparable weak satellites. Thus, the hydrocarbon of scotch tape has a broad signal at $I^* = 306$ eV and I = 311 eV representing the shake-up of the various MO constituted mainly (*165*) of carbon 2$p$.

## E. The Nickel(II) and Cobalt(II) Problem

A comparison (*15, 48*) of a large number of diamagnetic ($S = 0$) and paramagnetic ($S = 1$) $3d^8$ nickel(II) complexes show a striking difference in the Ni2$p$ region. The diamagnetic compounds show $2p_{1/2}$ and $2p_{3/2}$ as two normal, symmetric signals separated by 17.2 eV, whereas the paramagnetic compounds show four signals, the second and fourth situated mostly 5.8 to 6.2 eV above the first and the third signal separated by 17.4 to 17.9 eV. It is an argument in favour of the interpretation of all four signals as being due to levels of $2p^5 3d^8$ separated by effects of interelectronic repulsion (*48*) that this distance has increased, like it also does in $3d^3$ Cr(III) compared with Cr(VI) to 9.7 from 9.2 eV (*15, 130*) because the intrinsic spin-orbit coupling of inner shells is extremely insensitive to chemical bonding (*124*) and varies less than the uncertainty 0.1 eV in typical cases. However, it must be added in all fairness that measurements of NiO under high resolution (*151*) show a separation of 1.8 eV of two components of the first Ni2$p$ signal. Since we are not in the simple situation described by *Van Vleck* of ionizing an $s$ electron, it cannot be excluded that more components can be ascribed to the 270 states of $2p^5 3d^8$. On the other hand, *Kim* (*152*) finds a signal 7.2 eV above the lowest $I$(Ni3$d$) of NiO comparable to the satellite of $I$(Cu3$d$) in CuO, and ascribes it to monopole electron transfer between the bonding MO of symmetry type $e_g$ mainly constituted of O2$p$ to the anti-bonding sub-shell $e_g$ already containing two electrons in the groundstate of Ni(II), one in each of the orbitals having angular functions proportional to $(x^2 - y^2)$ and $(3z^2 - r^2)$. Though the ($S = 1$) examples discussed until now all are octahedral, essentially the same distribution of four signals in the 2$p$ region are found in tetrahedral nickel(II) complexes (*149, 166*). The decrease of the 6 eV distance by some 10 percent in compounds of reducing ligands (*149*) is compatible with a nephelauxetic effect, but seems too unimportant to suggest electron-transfer character.

It is an empirical fact that diamagnetic and paramagnetic nickel(II) compounds can be distinguished by the occurrence of two or four distinct signals in the Ni2$p$ region (and more rapidly than by a measurement of the magnetic susceptibility if the material has exclusively $S = 0$ or 1) but the detailed mechanism is still debated. Most authors believe that the second and the fourth signals of the ($S = 1$) cases are satellites of the copper(II) type, either of the electron transfer inter-atomic shake-up (*151, 152*) or intra-atomic (say $3d \rightarrow 4s$) shake-up (*146–150*) alternatives. In both cases, it is fairly difficult to justify the absence of satellites in diamagnetic ($S=0$) quadratic nickel(II) complexes having the strongly anti-bonding $3d(x^2-y^2)$ orbital empty and the four other $3d$-like orbitals fully occupied. The quantum number $L_V$ introduced by *Yin et al.* (*150*) has no physical significance. Actually, the groundstate of quadratic $d^8$ systems have a wavefunction which is a mixture of $^1D$, $^1G$ and $^1S$ components whereas the groundstate of octahedral $d^6$ systems is a definite mixture of $^1I$, $^1G$ and $^1S$ components. The only explanation for the absence of satellites in diamagnetic systems would be a numerical question of signal intensities. Seen from this point of view, it would be more attractive to ascribe the satellites to shake-up of the two anti-bonding electrons in the ($S=1$) nickel(II) complexes (*155*). An argument against electron transfer character is that the positions and the relative intensities of the second and fourth signal in the Ni$2p$ region do not seem to depend much on the reducing and covalent characteristics of the ligands in contrast to the lanthanum(III) satellites. However, the trimeric nickel(II) acetylacetonate containing the octahedral chromophores Ni(II)$O_6$ has unusually strong second and fourth signals which might be connected with the low-lying $\pi^*$ orbitals of this ligand, or with its unusually reducing character from the point of view of electron transfer spectra (*167*).

Before pursuing this discussion, it may be worthwhile to consider the other $3d$ group compounds. Diamagnetic ($S=0$) $3d^6$ cobalt(III) complexes have two straightforward $2p$ signals separated by 15.0 eV. It would appear that the weak undulation sometimes seen between these two signals, is due to superficial decomposition under the X-ray bombardment. On the other hand octahedral (*15*, *168*) and tetrahedral (*166*) high-spin ($S={}^3/_2$) cobalt(II) compounds have the two main signals separated by 16.0 eV. The broad structures correspond to two shoulders some 4 and some 6 eV above both $2p_{1/2}$ and $2p_{3/2}$. Hence, the typical ($S={}^3/_2$) Co(II) has six signals in the $2p$ region, which is not incompatible with the 720 states of the configuration $2p^5 3d^7$ (neglecting closed shells). *Briggs* and *Gibson* (*169*) studied two quadratic cobalt(II) complexes with $S={}^1/_2$. The two main signals are separated to the extent 15.4 eV, and the $2p_{3/2}$ has a weak shoulder 2 eV above the first $I$. This fascinating evidence is only compatible with interelectronic repulsion *or* with absence of anti-bonding electrons producing shake-up satellites.

The variation of the apparent spin-orbit coupling in the $2p$ shell is also pronounced in iron(II) complexes where diamagnetic ($S=0$) Fephen$_3^{+2}$ and Fe$(CN)_6^{-4}$ (*15*, *170*) have two signals separated by 12.9 eV whereas high-spin ($S=2$) iron(II) compounds such as the oxalate $FeC_2O_4$, $2H_2O$ have the distance 13.7 eV between the main signals having a shoulder at 3 eV higher $I$. Roughly the same distinction is found between low-spin ($S={}^1/_2$) iron(III) in Fe$(CN)_6^{-3}$ showing the distance 13.5 eV and high-spin ($S={}^5/_2$) Fe(III) a rather badly defined distance close to 14 eV combined with secondary signals 7 to 8 eV above the two main $I$-values. This is a rather different behaviour from the isoelectronic high-spin ($S={}^5/_2$) manganese(II) compounds having shoulders only some 4 eV above the $I$ of the two main signals separated by approximately 11.5 eV.

Several authors believe that Mn(II) shows genuine energy differences between $J$-levels of $2p^5 3d^5$ separated by differing interelectronic repulsion, whereas Fe(III), Co(II), Ni(II) and Cu(II) all show satellites of a different origin. If the hypothesis of delocalized anti-bonding electrons being a primordial condition for these satellites is accepted, one may invoke the much weaker nephelauxetic effect (*21, 34*) in Mn(II) than in Fe(III), and the absence of anti-bonding electrons in Cr(III) only showing shoulders (*15, 130*) some 2 eV above the $2p$ signals, to explain all the various observations. This is also the argument of *Wood* and *Urch* (*171*) finding complicated structure of the $1s3p^6 3d^5 \rightarrow 1s^2 3p^5 3d^5$ emission lines of Mn(II) and $1s3p^6 3d^3 \rightarrow 1s^2 3p^5 3d^3$ of Mn(IV) showing strong separations of the levels of the latter configurations, whereas $1s2p^6 3d^8 \rightarrow 1s^2 2p^5 3d^8$ emission lines of paramagnetic Ni(II) do not show such structure. The only weak point in this argument is that the selection rules for electric dipole transitions applicable to X-ray spectra conceivably may discriminate against certain levels. Such a situation is found in the relatively narrow groups (*172, 173*) of spectral lines of $4f$ group $M^{+2}$ and $M^{+3}$ in the ultra-violet due to transitions $4f^q 6p \rightarrow 4f^q 6s$ where each configuration may be 5 eV wide, but where the intense emission lines have unchanged parentage of the $4f^q$ part of the wave-function.

One has to ask the question whether the configuration $2p^5 3d^q$ (neglecting closed shells) is sufficiently wide to explain the controversial "satellites". This is rather difficult to know today. It is conceivable that the parameters of interelectronic repulsion between the $2p$ and the $3d$ shell are much smaller in the beginning, such as Cr(III) and Mn(II), than at the end of the iron group, such as Co(II) and Ni(II). Seen from the point of calculations, more reliable information is available for the $3s$ and $3p$ photo-electron signals corresponding to the configurations [Ne]$3s3p^6 3d^q$ and [Ne]$3s^2 3p^5 3d^q$, respectively. The $3s$ ionization is particularly favourable, because the spin-orbit coupling does not occur. Since the groundstate of $3d^3$Cr(III) belongs to the term $^4F$ in *Russell-Saunders* coupling, the two terms $^5F$ and $^3F$ are separated by a known linear combination of parameters of interelectronic repulsion, and corresponding to an experimental separation (*15, 130*) 2.3 eV. The two $3s$ signals of Mn(II) are essentially due to the terms $^7S$ and $^5S$ separated by 6.6 eV though higher terms can be perceived (*174*). The situation is rather different for nickel(II) where *Okusawa, Ishii* and *Sagawa* (*175*) find two additional signals about 6 and about 10 eV above the first, strongest $3s$ signal in NiO, $NiCl_2$ and $NiBr_2$. It is difficult to escape the conclusion that one of these signals is a satellite. *Bagus, Freeman* and *Sasaki* (*176*) find that strong effects of configuration mixing are needed to obtain agreement with the observed distance between the two components of the Mn$3s$ (6.5 eV in $MnF_2$ and 5.7 eV in MnO) and the Mn$2s$ (5.85 eV in $MnF_2$) signals.

A higher limit to the spin-orbit separation between $3p_{1/2}$ and $3p_{3/2}$ in the $3d$ group is the value 3 eV found (*15*) for Zn(II) and 3.4 eV for Ga(III). In the middle of the $3d$ group, much lower values are expected. *Combet Farnoux* (*177*) calculated (in intermediate coupling) the levels of $3p^5 3d^4$ (formed by photon absorption in $3p^6 3d^3$) of V(II) to be distributed between 37.6 and 50.0 eV and the levels of $3p^5 3d^8$ formed in Co(II) from 56.3 to 62.8 eV. The most intense transitions are calculated to occur at 43.1, 45.9 and 49.9 eV in V(II) and at 60.2 and 62.8 eV in Co(II). *Bagus* and *Wahlgren* (*178*) performed MO calculations on the octahedral chromophore Ni(II)$O_6$ finding quartet ($S = {}^3/_2$) levels (in *Russell-Saunders* coupling) at the energies (relative to the lowest level of the configuration $3p^5 3d^8$) 0, 5.0, 6.7 and 11.0 eV and doublet levels at 3.6, 8.7, 13.1, 14.6, 17.7 and 19.6 in good agreement with photo-electron results showing peaks at the relative energies

0, 2.3, 5.5, 7.5, 13 and 21 eV. Several calculations of $3s3p^6 3d^q$ and $3s^2 3p^5 3d^q$ term distributions have also been made by *Viinikka* and *Öhrn* (*179*).

It is clear that the comparable $\langle r \rangle$ of the $3p$ and $3d$ shells produce so large differences of interelectronic repulsion that the levels are distributed over an interval 10 to 20 eV broad, though the prominent photo-electron signals may be concentrated in a narrower range. For comparison, it may be mentioned that six (*26*) of the eight $(S,L)$-terms of the configuration $[Ar]3d^3$ are situated within 2.1 eV in $V^{+2}$ and within 2.6 eV in $Cr^{+3}$. It is not yet certain that $K_{av}(2p,3d)$ as a phenomenological parameter (*21*) is so much smaller than $K_{av}(3p,3d)$ and further on, the levels of $2p^5 3d^q$ are separated to a larger extent than given by the *Van Vleck* formula for $ns$ ionization. Hence, it is not sure whether the four or six signals in the $2p$ region of high-spin nickel(II) and cobalt(II) are due to a distribution of $J$-levels readily accessible (from the point of view of fractional parentage) by ionization. If the additional signals are due to shake-up (though it is noted that Mn(II) shows two $2s$ signals (*174*) separated by 6 eV) it is an intriguing question to what empty orbital the $3d$ electron is promoted. The oldest answer is the empty $4s$ orbital, but if the selection rule of monopole transitions remains important in this case, it may also be a kind of $4d$ shell with exactly the same symmetry types as the promoted $3d$ electron. This suggestion would suit energy differences between 6 and 10 eV fairly well. Since the major features of copper(II) are due to shake-up, it may also be that the observed spectra of Co(II) and Ni(II) are influenced both by shake-up and interelectronic repulsion (*273, 274, 276*). Calculations (*277*) on $2p$ in Ni(II) and (*278*) on $2p^5\ 3d^q$ show the latter effects slightly below 3 eV.

One way of resolving this problem would be to extrapolate from the observation of satellites at comparable distances from the main signals (*151*) of Cu$2s$, Cu$2p$, Cu$3s$, Cu$3p$ and Cu$3d$ that the M$1s$ signals (at very high $I$ and broadened by short half-life) also should show satellites, whereas $K(1s,3d)$ is expected to be very small. The only observation of $1s$ satellites reported (*153*) refers to $FeCl_3$ and $FeBr_3$ having satellites at 4.7 eV higher $I$. Unfortunately, this is exactly the case where one might suspect superficial reduction to Fe(II), and it would be highly valuable if the presence or absence of Ni$1s$ satellites in paramagnetic Ni(II) could be verified.

### F. Uranium Compounds

Metallic uranium (*180, 181*) has a photo-electron spectrum with distinct signals, the strongest originating in the $4f$, $4d$ and $5d$ shells. Most uranium(VI) compounds have similar spectra (*15, 37, 181*) with a modest chemical shift $dI$ amounting to 3 to 5 eV. This behaviour is comparable to the $d$ groups and very different from europium, where it was early shown (*39*) that metallic alloys containing the conditional oxidation state $4f^7$ Eu[II] and Eu(II) compounds such as $EuSO_4$ (*15*) have $I$ of all the inner shells 10 eV lower than of the Eu(III) compounds.

Nevertheless, various other uranium compounds may possibly exemplify all the known classes of satellites. We already mentioned that certain uranyl salts show shoulders of the U$4f_{5/2}$ and U$4f_{7/2}$ toward 3 eV higher $I$, which seem to be electron transfer satellites to the empty $5f$ shell in analogy to the satellites characterizing all lanthanum(III) compounds. *Keller* (*182*; *183*) prepared several mixed oxides known from magnetic and spectroscopic measurements to contain $5f^1$U(V) and in particular the fluorite types $MUO_4$ where the trivalent M = La, Gd, Ho and Lu. In all instances where all or most of the uranium is quinquevalent, one observes a broad and unusually intense satellite 8 eV to the left of U $4f_{5/2}$ (the distance 10.8 eV of this signal from U $4f_{7/2}$ makes it difficult to locate the exact position of the satellite belonging to the latter signal). It is not absolutely excluded that this satellite is due to electron transfer forming $5f^2$ in analogy to the cerium(III) compounds, but its high intensity and rather specific occurrence in the rare oxidation state U(V) makes it more probable that it is due to $5f \rightarrow 6d$ shake-up of the unique electron outside the closed shells (*186*).

The uranium(IV) double fluoride $K_2UF_6$ has nearly as high $I'(U\,4f_{7/2}) = 387.0$ eV as $RbUO_2(NO_3)_3$ (387.6) and $UO_2(IO_3)_2$ (387.5) whereas $U(SO_4)_2, 4H_2O$ has 386.4 eV and mixed oxides containing U(IV) this value marginally above 386 eV (*15*) to be compared with $I^* = 377.1$ or $I$ close to 382 eV for metallic uranium (*180*). Octahedral $UCl_6^{-2}$ exists in $[P(C_6H_5)_3H]_2UCl_6$ (for which $\partial$ could not be detected) showing $I(U\,4f_{7/2}) = 386.0$ eV. In all of these uranium(IV) compounds, satellites are seen of both U $4f$ signals at 5 to 7 eV higher $I$. This may again be due to shake-up (though the two $5f$ electrons in the groundstate are less delocalized than in the uranium(V) oxides) or, more likely, to interelectronic repulsion separating the $J$-levels of the configuration $4f^{13}5f^2$ (neglecting closed shells) containing $14.91 = 1274$ states. The twin signals cannot be due to superficial oxidation to U(VI) in view of the small chemical shift.

## G. Structure of Auger Spectra

The prototype of an *Auger* process is the decay of an *initial state* having a vacancy in an inner shell (say $1s$) by ejection of an electron having a kinetic energy $E_{kin}$ determined by the energy difference of the initial state and the *final state* presenting two vacancies in other shells (say $2s$ or $2p$). It is customary to write this transition $1s \rightarrow 2s2p$ or $1s \rightarrow 2p^2$ as a *motion of holes* or, even shorter but less obvious, $1s2s2p$ or $1s2p^2$ where it is a convention to indicate the initial vacancy at first. Seen from the point of view of atomic spectroscopy (*25*) there is no sharp distinction between Auger transitions and *auto-ionizing transitions* first discovered by *H. N. Russell* in the calcium atom having distinct levels belonging to the configuration [Ar]$3d4d$ at higher energy than the groundstate [Ar]$4s$ of $Ca^+$. It is one of the aspects of the classificatory validity of preponderant configurations (*21*) that the decay of these levels (technically situated in the continuum) by auto-ionization is comparatively slow. Less frequent *Auger* processes (*184*) involve the simultaneous emission of two electrons (dividing the energy in a fairly continuous way) or of one electron accompanied by shake-up [such as $1s2s^22p^63s \rightarrow$ ($1s^22s^22p^44s$ + free electron) in the sodium atom] or finally of initial states showing two or more vacancies.

The normal *Auger* spectrometer bombards metallic samples with an intense beam of (not necessarily monoenergetic) electrons, and the position of the signals is usually found by a device recording the differential quotient of the number of *Auger* electrons as a function of their $E_{kin}$. *Wagner* (*185*) pointed out that they can also be observed in favourable cases with a photo-electron spectrometer like our Varian IEE-15. This has the great advantage that non-conductors also can be studied (assuming that $E_{kin}$ has been decreased with the quasi-stationary potential represented by the C 1*s* signal difference $\partial$). Such samples acquire large, erratically varying negative potentials using electron bombardment. However, only certain elements show strong *Auger* signals, one of the conditions being that the initial ionization has $I$ as close as possible below $h\nu = 1486.6$ or 1253.6 eV. *Berthou* and the writer (*66, 192*) have obtained the *Auger* spectra of a large number of alloys and compounds using a photo-electron spectrometer. The chemical shift is a decrease of $E_{kin}$ normally, of the same order of magnitude as the increase of $I$. This can be qualitatively understood as a constant contribution $dI$ to the central field discouraging the ejection of the second electron. Two documents (*188, 189*) compile calculated *Auger* energies a → bc approximately equal to $I(\mathrm{a}) - I(\mathrm{b}) - I^{+}(\mathrm{c})$ where $I^{+}(\mathrm{c})$ of the loosest bound orbital is taken of the subsequent element with the atomic number $(Z+1)$ representing the effect of a vacancy in the orbital b.

Though this is a fairly good approximation to the barycentre of a given configuration, it is interesting to note that the configuration of the final state may be strongly split in individual energy levels. The classical case (*190*) is the initially ionized $Ne^{+}$ $1s2s^{2}2p^{6}$ producing *Auger* electrons with six different values of $E_{kin}$ forming the single state $^{1}S$ of $1s^{2}2p^{6}$, the two terms $^{1}P$ and $^{3}P$ of $1s^{2}2s2p^{5}$ and the three terms $^{1}S$, $^{1}D$ and $^{3}P$ of $1s^{2}2s^{2}2p^{4}$ also known (*26*) from the atomic spectra of $Ne^{+2}$. The same six terms are observed (*191*) in the cubic fluorides LiF and NaF, and in the isoelectronic (*187*) Na(I) and Mg(II). It is possible (*192*) to detect a chemical influence on the *Auger* spectra of fluoride in uniaxial ligand fields. The signal corresponding to $^{1}$S of $1s^{2}2s^{2}2p^{4}$ (which is weak in cubic crystals) intensifies and the distance between $^{1}S$ and the strong $^{1}D$ signal increases slightly, as one can predict from the asymptotic tendency of the former signal to correspond to a MO configuration lacking the two $\sigma$-bonding $2p$ electrons.

Argon (*58, 193*) shows $2p \rightarrow 3p^{2}$ hole transitions, which can also be observed in the isoelectronic K(I) and Ca(II). However, progressing from chromium to zinc, the $2p \rightarrow 3p3d$ and $2p \rightarrow 3d^{2}$ hole transitions become possible and more intense (*194, 195*). At this point, it is interesting to compare with krypton (*196*) having sharp *Auger* signals $2p \rightarrow 3d^{2}$ having the configuration [Ar]$3d^{8}4s^{2}4p^{6}$ of the final states. It is calculated that 8 levels occur between $^{3}F_{4}$ and $^{1}G_{4}$ in an interval 6.6 eV wide, and the single state $^{1}S_{0}$ at 8.4 eV above $^{1}G_{4}$. This is in good agreement with the *Auger* spectrum of krypton suggesting the $^{3}F$ levels 5.4 eV higher $E_{kin}$ than the sharp peak corresponding to $^{1}G_{4}$ whereas the isolated $^{1}S$ signal has 7.8 eV lower $E_{kin}$. Gaseous $Cu^{+3}$ (*197*) has the groundstate $^{3}F_{4}$ belonging to [Ar]$3d^{8}$ at 3.3 eV below $^{1}G_{4}$ again 4.3 eV below $^{1}S_{0}$. It turns out that metallic copper (*198–201*) and zinc (*199, 200*) have almost the same type *Auger* spectra as krypton atoms, with a total width of the $3d^{8}$ configuration 7 eV in Cu and 8 eV in Zn. This is a much more moderate nephelauxetic effect than known (*202, 203*) from the copper(III) complex $CuF_{6}^{-3}$. It may be noted that the configuration with two shells lacking an electron $3p^{5}3d^{9}$ is 15 eV wide in metallic copper (*198, 199*) and 25 eV wide in krypton (*196*). Linear copper(I) complexes have the *Auger* signals corresponding to the final configuration $3d^{8}$ somewhat modified (*192*) but not to the same extent as the broad *Auger* signals of Cu(II) complexes having final $3d^7$ states. The $^1G_4$ Auger signal of Cu is narrower than the photo-electron signals $I(\mathrm{Cu}\,2p)$ corresponding to the initial ionization, showing a longer half-life. Curiously enough, the initial ionization $2p_{1/2}$ produces *Auger* signals (*66, 187*) much less than half as intense as the initial $2p_{3/2}$ vacancy. Using 1253.6 eV photons in a photo-electron spectrometer, the *Auger* signals of Cu, Zn, Ga and Ge are almost as intense as the $2p$ photo-electron signals, whereas the Auger signals of F and Na are half as strong as the $1s$ signals. Using 1486.6 eV photons, the As and Se Auger signals are also very strong, and show spectacular chemical shifts (*15*). According to *Aksela* (*187*) zinc atoms have the $2p \rightarrow 3d^2$ hole transitions at 19 eV lower $E_{\mathrm{kin}}$ than in metallic zinc, suggesting strong relaxation effects in the metal. *Wagner* (*279*) recently showed the difference between the kinetic energies of an Auger electron and of the corresponding primary photo-electron (or between $I^*$ and the apparent $I^*$ of the Auger signal) to show chemical shifts (frequently 5 eV) almost exclusively) due to interatomic relaxation.

Xenon (*58, 196*) has strong $3d \rightarrow 4d^2$ hole transitions showing a width 10.3 eV of the configuration $[\mathrm{Kr}]4d^8 5s^2 5p^6$. The *Auger* spectrum is rather comparable with that of metallic silver (*204*) showing a width 4.6 eV (though *Schön* ascribes this separation to another reason). The elements from rhodium to lanthanum (*185—187*) show such $3d \rightarrow 4d^2$ *Auger* signals with a photo-electron spectrometer, which become supplemented (*66*) in the heavier lanthanides with $3d \rightarrow 4d4f$ and $3d \rightarrow 4f^2$ hole transitions. The last element to give such signals with 1253.6 eV photons is terbium, whereas erbium and thulium (*15*) produce strong *Auger* signals with 1486.6 eV photons, where the electrons have high $E_{\mathrm{kin}}$ imitating $I$ values around 104 and 46 eV, respectively. For $Z$ above 69, most *Auger* signals produced in photo-electron spectrometers are weak and uncharacteristic (*66, 185*) though an interesting theoretical aspect is that $E_{\mathrm{kin}}$ of the $4d \rightarrow 4f^2$ hole transitions *decrease* as a function of increasing $Z$ and are zero close to thorium. Moderate chemical shifts have been reported (*205*) for the *Auger* signals of uranium with $E_{\mathrm{kin}}$ below 300 eV.

It is evident that the continuous spectrum emitted by a synchrotron would be particularly helpful in studying *Auger* spectra, selecting photon energies close to the thresholds for the initial ionization. Such *synchrotron radiation* has already been used (*92*) in the region below 80 eV for obtaining photo-electron spectra. For our purpose, it is interesting to note the term separations amounting to 10 or 20 eV in a configuration where an inner shell lacks two electrons. The corresponding value of the spin-pairing energy parameter $D$ is 2.2 eV in $[\mathrm{Ar}]3d^8 4s^2 4p^6$ of $\mathrm{Kr}^{+2}$, or almost thrice the value in the lanthanides for the $4f$ shell. This is a complication usually forgotten by X-ray spectroscopists treating only one vacancy at the time.

## III. The Valence Region and Comparison with Studies of Gaseous Molecules

For the chemist, one of the most interesting consequences of the development of photo-electron spectrometry since 1962 is the experimental confirmation of the positions of penultimate MO in the valence region (*6, 206*), which can roughly be defined as $I$ below 40 eV though such as a definition cuts the fluorine 2s orbitals shortly, but include chemically inert 4f shells of lutetium, hafnium and tantalum. We restrict ourselves to specific problems of inorganic chemistry, because several contributors to the present volume treat the detailed MO distribution.

The comparative study of electron transfer spectra (*4, 105*) allowed the introduction of *optical electronegativities* $\chi_{opt}$ for ligands X and for central atoms M (varying as a function of the oxidation state). When discussing central atoms containing a partly filled $d$ or $f$ shell, it is not convenient here to introduce the corrections for the spin-paring energy Eq. (21) but rather define the uncorrected optical electronegativity $x_{uncorr}$ from the photon energy $h\nu$ of the first Laporte-allowed electron transfer band

$$h\nu = [x_{opt}(X) - x_{uncorr}(M)] \cdot 30000 \text{ cm}^{-1} . \tag{23}$$

The choice of the constant 30000 $cm^{-1}$ = 3.7 eV allows the halides $X^-$ to have $x_{opt}$ equal to the *Pauling electronegativity* 3.9 (F), 3.0 (Cl), 2.8 (Br) and 2.5 (I). It is possible to make a consistent set of optical electronegativities describing nearly all cases within a few thousand $cm^{-1}$ in spite of Eq. (20) though it must be admitted that one can run into *anisotropic complexes* (*207*) such as $UO_2I_4^{-2}$ or $NpO_2Cl_4^{-2}$ having long equatorial distances perpendicular on the linear uranyl and neptunyl ions with the result that $X^-$ seems less reducing than usual.

When comparing the lowest $I$ of numerous gaseous halides (*6*) it is striking that the values usually are within 1 eV from

$$\begin{gathered} I = (1 + 3.7\, x_{opt}) \text{ eV} \\ \text{F: } 15.4 \text{ eV} \quad \text{Cl: } 12.1 \text{ eV} \quad \text{Br: } 11.3 \text{ eV} \quad \text{I: } 10.2 \text{ eV} \end{gathered} \tag{24}$$

showing a rather unexpected connection with the electron transfer spectra. The values in Eq. (24) are slightly below the $I_1$ for gaseous halogen atoms in Eq. (19) but 12 to 7 eV above $I_0$ of gaseous $X^-$ representing the sum of the local *Madelung* potential and stabilization by covalent bonding (including the very fact of fractional atomic charge less negative than $-1$).

Tetrahedral molecules $MX_4$ show rather strong effects of inter-ligand repulsion separating the MO according to their number of angular node-planes "$l$" = 3, 2 and 1 for the three $\pi$ combinations formed from the eight $p\pi$ orbitals of the four X ligands. In the following, we give the "vertical" (*i.e.* unchanged internuclear distances) in eV:

| | $\pi t_1$ | $\pi t_2$ | $\pi e$ | |
|---|---|---|---|---|
| $CF_4$ | 16.3 | 17.5 | 18.6 | |
| $SiF_4$ | 16.46 | 17.54 | 19.5 | |
| $GeF_4$ | 16.1 | 17.0 | 18.5 | (25) |
| $CCl_4$ | 11.47 | 12.24 | 13.29 | |
| $TiCl_4$ | 11.78 | 12.78 | 13.23 | |

The unexpected result (*208*) that the carbon(IV) compounds are more reducing than the silicon(IV) compounds is due to the stronger X—X repulsion in the smaller $MX_4$. The effects of inter-ligand repulsion (*7, 105*) are rather important for the electron transfer spectra of tetrahalides such as $TiCl_4$, $FeCl_4^-$ and $CoBr_4^{-2}$ and of tetrachalkogenides (*209*) such as $MnO_4^-$ and $MoS_4^{-2}$. However, the total spreading of the excitation energies of $\pi$ orbitals in typical hexahalides such as $OsBr_6^{-2}$ and $IrCl_6^{-2}$ (*105, 210*) is about 1 eV, much less than the values in Eq. (25).

In the molecules $CF_4$ and $SF_6$, not only the $\sigma$ orbitals mainly constructed from F2*p* but also the deeper set of $\sigma$ orbitals corresponding energy-wise to F2*s* (*58, 124*) at $I=40.3$ ($t_2$) and 43.8 eV ($a_1$) in $CF_4$ and $I=39.3$ ($e_g$ with "$l$" $=2$), 41.2 ($t_{1u}$ with one angular node) and 44.2 eV ($a_{1g}$ totally symmetric) in $SF_6$ is separated by effects of chemical bonding. However, these exceptional cases have the F—F distance much shorter than twice the ionic radius of $F^-$, and the overlap integral (*124*) between two adjacent F2*s* orbitals is 0.11 in $CF_4$ and 0.05 in $SF_6$ to be compared with $5 \cdot 10^{-12}$ between two F1*s* orbitals in $CF_4$.

Diatomic halides were first studied as HX by *Lempka, Passmore* and *Price* (*211*) showing slightly higher $I$ than Eq. (24). It is possible by high-temperature equipment (*212—214*) to study diatomic alkaline halide molecules (or ion-pairs)

| MX = F | | Cl | Br | I | |
|---|---|---|---|---|---|
| H | 16.05, 20 | 12.74, 12.82, 16.6 | 11.67, 12.00, 15.6 | 10.38, 11.05, 14.25 | |
| Li | — | — | 10.0, 10.5, 11.4 | — | |
| Na | — | 9.0, (9.5) | 8.7, 9.4, 10.3 | 8.0, 9.0, 9.9 | (26) |
| K | — | 8.7, 9.5 | 8.1, 8.6, 9.3 | 7.4, 8.4, (9.2) | |
| Rb | — | 8.7, 9.4 | 8.0, 8.4, (8.8) | 7.3, 8.2, — | |
| Cs | 9.7, (10.1) | 8.5, (9.1) | 8.0, 8.4, (8.8) | 7.6, 8.5, — | |

It is an experimental difficulty that such vapours frequently contain a large proportion of rhombic dimers $XM_2X$. The monomeric TlF has the lowest $I=10.8$ eV for a 6*s*-like orbital (the first fluoride orbitals have $I=11.9$ eV) and forms a linear dimer FTlTlF (rather similar to solid calomel ClHgHgCl) with the lowest $I=9.95$ eV (*215*). The Tl6*s* orbital (perhaps concentrating on the rear side of the molecule (*216*) by mixing with the rotationally, symmetric Tl6*p*(*z*) orbital) in TlCl has $I=9.8$ eV. In such measurements, internal references of $I$ are provided by the line signals of admixed noble gases (*6*) or by the rather inevitable traces of $H_2O$ having a sharp signal at $I=12.61$ eV from the non-bonding orbital. The vertical $I$ in Eq. (26) refer to $\pi$ and $\sigma$ orbitals in fluorides and chlorides, and to the two spin-orbit components ($\omega = 3/2$ and $1/2$ of $\pi$ and the $\sigma$ orbital in HCl, bromides and iodides. It is noted that the separation of the two $\omega$-components in HBr(0.33 eV) and HI(0.67 eV) are about two-thirds of the free atoms Br(0.46) and I(0.95) in agreement with a simple description of $\pi$ orbitals, whereas the splitting is almost atomic in the monomeric alkaline halides. Correspondingly, the separation (*212—214*) between the diagonal elements of $\pi$ and $\sigma$ energy is rather small, below 0.6 eV, with strong mixing of $\pi$ and $\sigma$ character (*217, 218*) and restoring the approximate spherical symmetry of the outermost $p$-shell of $X^-$. This separation is estimated (*105, 219*) to 1.5 eV in hexachloride and 1 eV in hexaiodide complexes (it is less certain in fluorides) from the excitation energies of electron transfer spectra. Gaseous CsF has signals (*212*) at 18.3 and 19.1 eV with a rather irregular shape, which can be ascribed to $Cs5p_{3/2}$ and $Cs5p_{1/2}$ (situated at 1.72 eV above the groundstate of $Cs^+$). It is seen that the lowest $I$ values of the alkaline halides in Eq. (26) are considerably smaller than in polyatomic halides described by Eqs. (24) and (25). This is in qualitative agreement with the *Madelung* energy of an ion-pair, because the *Madelung* constant $\alpha$ is only 1, though the internuclear distance $R$ is shorter. One would expect the lowest $I$ value of the sodium halides to be $3.61 + (14.4/2.36) = 9.7$ eV for NaCl, $3.36 + (14.4/2.50) = 9.2$ eV for NaBr and $3.06 + (14.4/2.71) = 8.4$ eV for NaI. It is seen that the variation of Eq. (26) with the cation and with the anion is roughly the same as the $I'$ of crystalline alkaline halides given in Table 1.

The experimental effort on the fairly corrosive inorganic halides was somewhat reluctant until the Oxford group invented a cheap, expendable metal cylinder for the 21.2 eV spectrometer. The first 3$d$-like electron detected in a gaseous halide was a weak, sharp signal at 9.41 eV in $VCl_4$ (*220*) having in addition almost the same $I$-values as the $TiCl_4$ shown in Eq. (*25*). Table 3 gives a list of $I$ for $d$-like electrons found in gaseous molecules (*220—231*) and for comparison, $I'$ and $I$ values in solid compounds (*15*). For a given oxidation state, the lowest $I$ of each gaseous compound varies quite a lot, about 2 eV between complexes of structurally so similar ligands as acetylacetonate and hexafluoro acetylacetonate (a remote analogy to the "perfluoro effect" (*232*) recognizing $\sigma$-bonding orbitals of hydrocarbons by their strongly increased $I$ by fluorine substitution of hydrogen atoms) and 3 eV from the sandwich chromium(0) complex $Cr(C_6H_6)_2$ (with $N = 12$) to $Cr(CO)_6$. The lowest $I$ known for a gaseous molecule is 5.0 eV for the corresponding mesitylene complex (to be compared with $I$ of solid alkaline metals close to 2 eV, or the gaseous caesium atom 3.89 eV) whereas $Cr(PF_3)_6$ has $I = 9.0$ eV. This chemical shift may be due to varying fractional charge, to bonding of the $d$-orbitals to empty orbitals of the ligands (traditionally called "back-bonding") or a combination of both.

It is seen from Table 3 that the surrounding ligands are at least as important for the variation of the $I$ values as the oxidation state in an isoelectronic series. This forms a striking contrast to the lanthanides (*28*) where $I(M4f)$ increases 8 eV per unit of oxidation state in isoelectronic M(II), M(III) and M(IV). This is almost as large a shift as 9 eV found between Lu(III), Hf(IV), Ta(V), W(VI), . . . up to gold. In the closed-shell $d^{10}$ systems, the same steep variation with oxidation state is found in Cu(I), Zn(II), Ga(III), . . . or Ag(I), Cd(II), In(III), . . . but only for the positive oxidation states. The $I$ of two different sub-shells can be found in gaseous nickel(0) compounds, $\Delta$ (being negative (*7, 13*) in tetrahedral chromophores) is $-0.87$ eV in $Ni(CO)_4$ and $-1.03$ eV in $Ni(PF_3)_4$ though the value of $\Delta$ obtained by internal transitions in the partly filled shell may not be strictly identical (*21, 233*) with the difference between $I$ of the two sub-shells. In this connection, it is worth mentioning the measurements (*234*) of crystalline CuCl, CuBr and CuI showing two $I(3d)$ separated by about 1 eV, below the halogen

Table 3. Ionization energies of *d*-like electrons in gaseous transition-group compounds and $I'$(corrected) or $I$(uncorrected for charging effects) of solids (*15*). The values marked with an asterisk are higher than the ionization energy of the loosest bound MO mainly situated on the ligands. All in the unit eV. The ligands are abbreviated aca⁻ = acetylacetonate. hfa⁻ = hexafluoroacetylacetonate and mes = 1.3,5-trimethylbenzene

| | | |
|---|---|---|
| $3d^1$ | Ti(III) | Ti $hfa_3$ 7.94 (*221*) |
| | V(IV) | $VCl_4$ 9.41 (*220*) |
| $3d^2$ | V(III) | V $hfa_3$ 8.68 (*221*) |
| $3d^3$ | V(II) | $V(C_5H_5)_2$ 6.78 (*227*) |
| | Cr(III) | Cr $aca_3$ 7.5 (*221*) Cr$hfa_3$ 9.57 (*221*) $Cr_2O_3$ $I = 8.2$ |
| $3d^5$ | V(0) | $Vmes_2$ 5.33, 5.61, 6.08 (*223*) $V(CO)_6$ 7.52, 7.88, (8.4) (*222*) |
| | Cr(I) | $Cr(C_5H_5)(C_6H_6)$ 6.20, 7.15 (*223*) |
| | Mn(II) | $Mn(C_5H_5)_2$ ($S = 5/2$) 6.91, 10.10*, 10.51* (*227*) |
| | Fe(III) | Fe $hfa_3$ 11* (*221*) $Fe_2O_3$ $I = 10.6$, 13.6* $(NH_4)_2[Fe(H_2O)F_5]$ $I = 12.1$, 13.6* |
| $3d^6$ | Cr(0) | Cr $mes_2$ 50.1, 5.88 (*223*) $Cr(C_6H_6)_2$ 5.4, 6.4 (*223*) $Cr(CO)_6$ 8.40 (*224*) $Cr(PF_3)_6$ 9.0 (*225*) |
| | Mn(I) | $Mn(C_5H_5)(C_6H_6)$ 6.36, 6.72 (*223*) $Mn(CO)_5H$ 9.0, 10.6 (*226*) $Mn(CO)_5Cl$ 10.46*, 11.08* (*226*) |
| | Fe(II) | $Fe(CN)_6^{-4}$ $I = 7$ Fe $phen_3^{+2}$ $I = 8$ $Fe(C_5H_5)_2$ 6.88, 7.23 (*227*) |
| | Co(III) | Co $aca_3$ 8 (*221*) Co $hfa_3$ 10.1* (*221*) $[Co(NH_3)_5Cl]Cl_2$ $I = 9.2$ $Li_3[Co(CN)_6]$ $I = 9.6$ |
| $3d^7$ | Co(II) | $Co(C_5H_5)_2$ 5.56, 7.18, 7.63 (*227*) $Co_4aca_8$ $I = 7.4$ $CoF_2$ $I' = 10$ |
| $3d^8$ | Fe(0) | $Fe(CO)_5$ 8.6, 8.9 (*225*) $Fe(PF_3)_5$ 8.9, 10.2 (*225*) |
| | Ni(II) | $Ni(C_5H_5)_2$ 6.51, 8.43 (*227*) NiO $I' = 7.7$ $KNiF_3$ $I' = 9.3$ |
| $3d^9$ | Cu(II) | $CuF_2, 2H_2O$ $I' = 11.1$ |
| $3d^{10}$ | Fe(—II) | $Fe(NO)_2(CO)_2$ 8.56, 8.97, 9.74 (*228*) |
| | Co(—I) | $Co(NO)(CO)_3$ 8.90, 9.82 (*228*) |
| | Ni(0) | $Ni(CO)_4$ 8.90, 9.77 (*228*) $Ni(PF_3)_4$ 9.55, 10.58 (*229*) |
| | Cu(I) | CuCN $I' = 8.9$ |
| | Zn(II) | $Zn(CH_3)_2$ 16.96* (*231*) $ZnI_2$ 18.39* (*230*) $ZnBr_2$ 18.69* (*230*) $ZnCl_2$ 19.02* (*230*) $[Zn(H_2O)_6](SiF_6)_2$ $I' = 16.5$* $Zn(IO_3)_2$ $I' = 15.6$* |
| | Ga(III) | $Ga_2O_3$ $I' = 25.6$* $GaF_3, 3H_2O$ $I' = 26.8$* |
| $4d^5$ | Ru(III) | Ru $hfa_3$ 8.85, 9.07 (*221*) $[Ru(NH_3)_6]Cl_3$ $I = 8.9$ |
| $4d^6$ | Mo(0) | Mo $mes_2$ 5.13, 6.03 (*223*) $Mo(C_6H_6)_2$ 5.32, 6.33 (*223*) $Mo(CO)_6$ 8.50 (*224*) |
| | Rh(III) | $RhCl_3$ $I = 8.9$ $Na_3RhCl_6, 12H_2O$ $I = 9.5$ $[Rh(NH_3)_5Cl]Cl_2$ $I = 9.6$ |
| $4d^8$ | Rh(I) | $RhH(PF_3)_4$ 9.7, 11.8 (*225*) |
| | Pd(II) | $[Pd(NH_3)_4](CH_3C_6H_4SO_3)_2$ $I = 9.7$ $K_2PdCl_4$ $I = 10.2$ |
| $4d^{10}$ | Ag(I) | AgBr $I' = 10.0$ $AgIO_3$ $I' = 10.5$ |
| | Cd(II) | CdO $I' = 16.8$ $CdI_2$ $I' = 17.3$ |
| | In(III) | $In_2O_3$ $I = 23.0$ $In_2(SO_4)_3$ $I' = 24.3$ $In(IO_3)_3$ $I' = 24.1$ |
| $5d^3$ | Re(IV) | $K_2ReCl_6$ $I' = (8)$, 9.7 |
| $5d^4$ | Os(IV) | $K_2OsCl_6$ $I' = (8.6)$, 10.6 |
| $5d^6$ | W(0) | $W(CO)_6$ 8.56 (*224*) |
| | Ir(III) | $I' = 9$ |
| | Pt(IV) | $Rb_2PtCl_6$ $I' = 9.6$ $Cs_2PtCl_6$ $I' = 9.7$ |
| $5d^8$ | Pt(II) | $[N(C_4H_9)_4]_2PtCl_4$ $I = 8.3$ $K_2PtCl_4$ $I = 9$ |
| $5d^{10}$ | Pt(0) | $Pt(PF_3)_4$ 9.8, 12.5 (*225*) |
| | Hg(II) | $Hg(CH_3)_2$ 14.95, 15.44, 16.9* (*231*) $HgI_2$ 15.99, 16.17, 17.91* (*231*) $HgBr_2$ 16.40, 16.69, 16.83, 18.34* (*231*) $HgCl_2$ 16.71, 17.05, 17.27, 18.65* (*231*) HgO $I' = 14.0$, 15.4* $Hg(IO_3)_2$ $I' = 14.1$, 15.5* $HgF_2$ $I' = 14.3$, 15.7* |

*np*-signal some 3 to 2 eV higher *I*. We find $I'(\text{Cu}\,3d)$ of CuCN approximately 2 eV above these halides. $\Delta$ is $-0.92$ eV in $Co(NO)(CO)_3$ being a cobalt($-$I) complex isoelectronic with $Co(CO)_4^-$ if the ligand is $NO^+$. The moderate sub-shell energy differences 0.3 eV in $Fe(CO)_5$ and in $Fe(C_5H_5)_2$ correspond to roughly non-bonding orbitals, whereas the differences are 1.0 eV in $Cr(C_6H_6)_2$ and 1.3 eV in $Fe(PF_3)_5$.

It is only in diamagnetic groundstates with $S=0$ that one can hope to identify individual sub-shells by a single signal. It was first pointed out (*222*) in the case of low-spin ($S={}^1/_2$) $3d^5$ $V(CO)_6$ that ionization of one of the five electrons in the lower ($xy,xz,yz$) sub-shell produces three different levels (one triplet and two singlets) of $3d^4$. The visible absorption spectrum (*235*) leaves some suspicion that this molecule might be *Jahn-Teller* distorted on an instantaneous picture (*7*) but the general principle is valid that the relative probability of forming various final states of the configuration $(nl)^{q-1}$ is proportional to the squares of the coefficients of fractional parentage in the groundstate, as discussed by *Cox* in this volume. The situation is fairly complicated in $3d^q$ compounds with positive *S*. *Wertheim*, *Guggenheim* and *Hüfner* (*236*) find a shoulder at lowest *I* in the ($S=2$) $FeF_2$ forming the groundstate ${}^6A_{1g}$ of Fe(III) whereas the two other signals are due to quartet levels, partly coinciding with the F $2p$ signal. Octahedral cobalt(II) compounds with $S={}^3/_2$ have a broad, rather undistinct distribution of $I(\text{Co}\,3d)$ of which the first signal is expected to be due to ${}^5T_{2g}$ followed by a series of triplet states. NiO and $NiF_2$ (crystallizing in rutile type) containing the octahedral chromophores $Ni(II)O_6$ and $Ni(II)F_6$ show a sharp signal due to ${}^4T_{1g}$ followed by ${}^2E_g$ and ${}^2T_{1g}$ or ${}^2T_{2g}$ (*236*). It is characteristic for these three cases that the lowest $I(3d)$ corresponds to a removal of an electron from the *lower* sub-shell. This apparent paradox is explained by the strong effects of interelectronic repulsion, compensating the energy $2\,\Delta$ needed for the promotion of two electrons to the higher sub-shell by the favourable spin-pairing energy in Eq. (21) for the highest *S* possible. Actually, the very existence of high-spin complexes (able to become low-spin for a sufficiently large ratio $\Delta/D$) is explained the same way. Since the variation of $\Delta$ is intimately connected with decreasing internuclear distances (*7*) it is probable that the distribution of final $3d^{q-1}$ states corresponds to roughly the same $\Delta$ as in the groundstate, because the photo-electron spectra obey the *Franck-Condon* principle. On the other hand, the groundstate ${}^4A_{2g}$ of $Cr(III)O_6$ in $Cr_2O_3$ shows only one signal (*15, 130*) because only the level ${}^3T_{1g}$ containing two electrons in the lower sub-shell can be formed. In sofar diamagnetic $d^6$ complexes are close to the pure sub-shell configuration with the lower sub-shell filled, only one signal is expected, corresponding to the final state ${}^2T_{2g}$. One would expect high-spin ($S={}^5/_2$) $d^5$ systems to be more informative, having two signals (corresponding to ${}^5T_{2g}$ and ${}^5E_g$) separated by $\Delta$. However, the practical difficulty is the possible coincidence with signals due to ligand orbitals in our two iron(III) compounds. The separation 1.5 eV in the aqua-fluoro complex corresponds the best to the spectroscopic $\Delta = 1.8$ eV.

We are not going here to discuss the *d*-like energy bands of metallic transition elements except for two pioneer studies (*237, 238*) and an important demonstration (*239*) that palladium-silver alloys behave as if individual $I(\text{Pd}\,4d)$ and $I(\text{Ag}\,4d)$ remain distinct. *Watson* and *Perlman* discuss gold and its alloys in great detail in the present volume. Many authors (*240—245*) have studied the valence region of semiconducting elements such as Si, Ge, Se and Te and of semiconducting compounds such as ZnS, ZnSe, ZnTe, GaP, GaAs, GaSb, CdS, CdSe, CdTe, InP, InAs, InSb, PbS, PbSe and PbTe. For the purpose of detecting the weak details in the valence region, it is particularly helpful (*245*) to use monochromatized photons. It is of general chemical interest that three signals of diamond (*241*) occur close to $I = 21$, 16 and 10 eV to be compared with a measurement (*165*) of the alkane $C_{36}H_{74}$ and a calculated band structure of polyethylene $(CH_2)_n$. Diamond (*241*) has a shake-up signal at $I = 324$ eV. This position 34 eV above the 25 times stronger, main C1*s* signal has no relation to a lower *I*-value, but demonstrates a changed Hartree potential. Gaseous $CH_4$ has $I = 14.5$ and 23.08 eV for the $\sigma$ type $t_2$ and $a_1$ orbitals (*241*) to be compared with 12.67 and 18.01 eV for $SiH_4$ and 12.28 and 18.46 eV for $GeH_4$.

It is frequently believed by solid-state physicists that the photo-electron spectra of the valence region indicate the *density* of energy band states as a function of the ionization energy $I^*$ relative to the *Fermi* level. This statement is open to caution, because the signal intensities discriminate strongly in favour of *d* and *f* electrons. This discrimination increases strongly with increasing photon energy, as first demonstrated by *Eastman* and *Kuznietz* (*246*) in the case of the NaCl type semiconductor EuS showing stronger S3*p* than Eu4*f* signals at $h\nu = 20$ eV but much stronger Eu4*f* signals for $h\nu$ above 40 eV. Pyrite $FeS_2$ shows the same phenomenon (*247*) with S3*p* signals at $I^* = 5$ and 8 eV, whereas a sharp Fe3*d* peak at $I^* = 1$ eV progressively becomes much stronger with increasing photon energy. By the same token, the metallic compound $ReO_3$ (crystallizing in the cubic perovskite type with the $N = 12$ positions empty) show a signal (*248*) at $I^* = 0.5$ eV due almost exclusively to the Re5*d* component of the MO because oxygen 2*p* gives a much weaker signal (*118*) per electron. It is the high intensity produced by each of the three to six 5*d*-like electrons in the lower sub-shell of $MCl_6^{-2}$ (M = Re, Os, Ir, Pt) which makes it possible to distinguish the signals (*15*) given in Table 3. Comparable observations including $SnCl_6^{-2}$ were made by *L. E. Cox* and *Hercules* (*249*).

We are not going here to discuss at length the influence of substitution in the valence region *I* values. The well-known ligand $NH_3$ has the vertical $I = 10.85$ eV to be compared with $I = 10.6$ eV for the lone-pair of $PH_3$ (*6*). *Bassett* and *Lloyd* (*250*) find an increase in the fluorides to 13.7 in $NF_3$ and 12.3 eV in $PF_3$ not much below the lowest *I* in the tetrahedral molecules 14.1 eV in $ONF_3$ and 13.5 eV in $OPF_3$. Methyl substitution has the opposite effect (*251*), the four $M(CH_3)_3$ have almost identical *I* for the lone-pairs 8.54 (M = N), 8.60 (P), 8.65 (As) and 8.48 eV (M = Sb).

We return in Chapter IVB to the question that all the "vertical" *I* values are higher than ionization energies related to thermochemical equilibria, where the internuclear distances may re-arrange to a large extent. By the same token as a broad electron transfer band at a given wave-number (*105*) is more liable to indicate a tendency toward spontaneous reduction than a narrow band, it is also true that broad photo-electron signals for a given *I* correspond to a more reducing molecule in the chemical sense than narrow signals. Salts of a given cation are also stabilized by the Madelung potential relative to *I* for forming the cation by ionization of the molecule in the gaseous phase.

## IV. Previous Ideas of Bonding in Solids

### A. Hybridization in VB and *l*–Mixing in LCAO

The photo-electron spectra, in particular in the valence region for $I$ below 40 eV, have provided much information of direct or indirect relevance to the theory of chemical bonding. It must be said in all fairness that spectroscopic studies of optical excitations, and *a fortiori* ionization, intrinsically emphasize the MO description, whereas there exists a quite legitimate, rather orthogonal description exclusively interested in the groundstate, where the angular dependence of chemical bonding (which is not excessive, as seen from the existence of $P_4$ and cyclopropane with the unusually low bond angles 60°) and bond dissociation energies (which are difficult to incorporate in a quantum-mechanical description adapted to treat one set of internuclear distances at a time) are of importance. Nevertheless, it must be realized once for all that one of the major assumptions of the VB (valence-bond) theory (that $N$ orbitals are involved in the chemical bonding in a chromophore $MX_N$ with $N$ equivalent ligating atoms) cannot be generally valid (*13, 252*). NaCl type oxides, nitrides and carbides do not possess six valence orbitals of the anion, the molecules $SF_6$ and $IrF_6$ hardly have six equivalent central atom orbitals, and the whole *d*-group spectroscopy is against this idea.

It is true that the $2s$ and $2p$ radial functions of the elements from beryllium to fluorine are sufficiently similar that one may, to the first approximation, put a common radial function outside parenthesis multiplied with a linear combination (a *hybridization*) of the four angular functions $1, (x/r), (y/r)$ and $(z/r)$. Nevertheless, the fact (*6, 241*) that the C$2s$-like orbital of methane has $I$ 70 percent higher than the three C$2p$-like orbitals is a prototype of a situation prevailing in all carbon, nitrogen, oxygen and fluorine compounds, where one can recognize *energetically* the $2s$-like MO in the region above $I = 20$ eV quite distinct from the lower $I$-values. However, the situation is entirely different for the elements beyond magnesium. It would be very difficult to point to any comparable radial function for six orbitals in $AlF_6^{-3}$, $SiF_6^{-2}$ and $PCl_6^-$. In the $3d$ group, the average radii have the ratio close to 1:3:4 for $3d, 4s$ and $4p$. Still, it has been necessary to discuss in detail (*253*) a suggestion by *Pauling* and *Keaveny* for rescuing the $d^2sp^3$ hybridization.

In MO theory, one has another kind of hybridization which it is convenient (*7*) to call *l-mixing* in order to avoid confusion. Seen from a strictly group-theoretical point of view, orbitals of a given atom may mix if the have they same symmetry type in the local point-group. Hence, if the nuclear skeleton possess no elements of symmetry, all orbitals are capable of mixing in the lowest point-group $C_1$. It is obvious that the interesting question is whether the mixing is numerically important. The gaseous diatomic molecules have the local symmetry $C_{\infty v}$ at the two nuclei, and the mixing of $ns$, and $np(z)$ orbitals is enhanced (*216*) by the empty space available at the rear side of the nucleus. Evidence is available that the $\sigma$ orbitals in halides $MX_N$ and the corresponding complex anions consist of 10 to 20 percent of $ns$ mixed into $np$, and *vice versa*. On the whole, the $l$ mixing may not be as important in occupied MO as one might expect. The situation is rather similar with the two-electron operator (*4, 21*) of interelectronic repulsion capable of mixing terms with the same combination of $S$ and $L$ (assuming *Russell-*

*Saunders* coupling) and the same parity in a given monatomic entity. Since there is a transfinite number of configurations containing such terms, the problem is whether a given configuration has a perceptible or a negligible influence. At the time around 1930 when atomic spectroscopists were numerous (just before nuclear transmutations became fashionable) the general consensus was that configuration interaction is important when two configurations of the same parity overlap, or at least are adjacent with a distance comparable to the width of the widest configuration. However, it has turned out (*4*) that two-electron substitutions into orbitals situated in the continuum are far more important in the majority of cases. By the same token, the deviation from the minimum LCAO approach deforming the atomic orbitals by a minor admixture of continuum orbitals may explain a variety of chemical facts (*79*).

## B. "Ligand Field" Theory

The energy levels of *d* and *f* group complexes can be described (*7, 21*) as a moderate perturbation of the ligating atoms on the five *d*-like or seven *f*-like orbitals constituting some of the MO of the complex considered. Whereas this perturbation originally was believed (*44, 45*) to consist of (or at least be qualitatively represented) by the non-spherical (small) part of the *Madelung* potential, it is more plausible that the anti-bonding character of some of the *d*-like or *f*-like orbitals (*252*) is adequately described by the *angular overlap model* (*7, 254*) which can be shown to be equiconsequential with a contact potential acting at the ligand positions.

Photo-electron spectra have been extraordinarily uninformative about the sub-shell energy differences described by "ligand field" theory because the only d group complexes containing anti-bonding electrons (excluding the $d^{10}$ systems in Table 3) are exactly those where the lowest ionization energy is shown by an electron in the lower, roughly non-bonding sub-shell, as discussed in Section III. Said in other words, if one succeeded in measuring the 3d region with very high resolution of octahedral high-spin Fe(II), Co(II) and Ni(II) one would see a distribution of energy levels very similar to that known (*13*) for the preceeding element Mn(II), Fe(II) and Co(II) where the lowest level still contains two electrons in the higher anti-bonding sub-shell consisting of $(x^2-y^2)$ and $(3z^2-r^2)$.

Nevertheless, photo-electron spectra have had a great impact on "ligand field" theory at one point, colloquially called (*2, 31*) the "third revolution". The *Hückel* model of a homonuclear system is essentially a topological matrix if one has one kind of non-diagonal element $\beta$ between neighbour atoms, and a fixed diagonal element $\alpha$ of the effective one-electron operator. Mildly heteropolar molecules (such as pyridine) are normally described by $\alpha_N$ being somewhat more negative than $\alpha_C$. A variety of related models for inorganic molecules and polyatomic ions can be proposed. Around 1960, the *Wolfsberg-Helmholz* model (*4*) was quite popular among transition-group spectroscopists. Its results can be summarized that with $0 > \alpha_M > \alpha_X$ the eigen-value corresponding to the antibonding orbitals in the partly filled shell of $MX_N$ is approximately

$$E = \alpha_M + k(\alpha_M)^2(S_{MX})^2/(\alpha_M - \alpha_X) \tag{27}$$

with $k$ close to 1, and the overlap integral $S_{MX}$ between the M orbital considered and a symmetry-adapted linear combination of X orbitals having the same symmetry type.

We have held this discussion as general and qualitative as possible, in order not to get into the (obviously debatable) details of each model *(4)*. It is characteristic for all *Hückel-Wolfsberg-Helmholz* types of models that the delocalization of the anti-bonding orbital becomes more pronounced when the denominator $(\alpha_M - \alpha_X)$ decreases. In the case of $\alpha_M = \alpha_X$ one would replace Eq. (27) by

$$E_1, E_2 = \alpha_M \pm k\, \alpha_M\, S_{MX} \tag{28}$$

in the first approximation (where $k$ would be slightly larger for the higher $E_2$ with the minus sign than for the bonding $E_1$ with the plus sign to the negative $\alpha_M$). It has been discussed *(7)* based on a suggestion by *Ruedenberg* *(24)* that the non-diagonal $\alpha_M$ in Eqs. (27) and (28) may represent the local contribution to the kinetic energy $\mathscr{P}$ of Eq. (10) in the bond region between the M and X nuclei. Anyhow, most authors believed implicitly that the diagonal elements $\alpha_M$ and $\alpha_X$ represent uncorrected ionization energies of the unperturbed atomic orbitals, though one might also argue that they represent the *Mulliken electro negativity*, the average of the electron affinity and the ionization energy of the orbital *(4)*.

This way of thought has got a serious stroke from the observation *(221)* that the $3d$-like electrons of iron(III) and cobalt(III) tris (hexafluoroacetylacetonate) have *higher I* than the loosest bound MO distributed on the ligands. Whereas the $3d^6$ Co(III) has all six electrons in the lower sub-shell of the approximately octahedral chromophore $M(III)O_6$ the situation is far more paradoxical in the high-spin ($S = 5/2$) Fe(III) having one electron in each of five orbitals constituted by the lower sub-shell as well as the anti-bonding sub-shell with the astounding result that an anti-bonding MO may have a higher $I$ than its bonding counterpart with the same symmetry type. Actually, higher $I$ of d-like orbitals (but lower electron affinity) than of the loosest bound MO had been contemplated *(4)* for combinations of oxidizing central atoms and reducing ligands such as $OsI_6^{-2}$ and $CuBr_4^{-2}$ with low wave-numbers of the electron transfer bands. A famous case of this inversion in a careful MO calculation is $Ni(CN)_4^{-2}$ studied by *Demuynck* and *Veillard* *(255)* predicting eight MO (mainly located on the cyanide ligands) with $I$ between 3.29 and 4.35 eV followed by the $3d$-like orbitals, $(3z^2 - r^2)$ at 4.95, $(xz)$ and $(yz)$ at 7.26 and $(xy)$ at 9.1 eV. It is remarkable that the first spin-allowed transition $(3z^2 - r^2) \rightarrow (x^2 - y^2)$ is displaced by effects of Eq. (20) to 2.56 eV in reasonable agreement with the experimental value 3.4 eV. It can be seen from the resistance of the complex toward oxidation in aqueous solution that $I$ of the loosest bound MO must be above 6 eV, whereas solid $K_2Ni(CN)_4$ shows $I(3d) = 9$ eV. Unfortunately, this sample *(15)* is not a typical non-conductor, and the charging effect $\partial$ may be somewhere between O and 2 eV providing a lower limit $I'(3d) = 7$ eV. It is typical for MO calculations on polyatomic anions that the background *Madelung* stabilization is neglected or underestimated. For comparison, it may be noted *(255)* that the tetrahedral molecule $Ni(CO)_4$ has the lowest three $I$ values calculated to be 10.75, 12.82 and 17.52 eV where the observed "vertical" $I$ are 8.90, 9.77 and 14.8 eV (cf. Table 3) and the lowest $I$ of CO is 14.01 eV (*6*).

Photo-electron measurements of $5d^8$ $K_2Pt(CN)_4$ show (*256*) signals at $I^* = 3.0$, 5.0 and 8.1eV and of $5d^6$ $K_2[Pt(CN)_4Cl_2]$ 4.4 and 7.8 eV (the $I$ values relative to *vacuo* are some 5 eV higher). These values have been compared (*256*) with the compound $K_2[Pt(CN)_4Cl_{0.3}]$ which is a one-dimensional metal in the same sense as graphite is a two-dimensional metal. *Van der Lugt* (*257*) compared MO clacula-tions of $PdCl_4^{2-}$ and related species with photo-electron spectra. Like other chloro complexes of noble metal central atoms (*15, 249*) the *d*-like orbitals (strongly intensified (*118*) signals) show slightly lower $I$ than Cl3*p*. Inversions with higher $I$(Cu3*d*) than of the halide orbitals are also found in the (rather hypothetical) planar $CuF_4^{-2}$ (*258*) and $CuCl_4^{2-}$ (*259*). A rather extreme case is the planar mol-ecule $FeF_3$ with $S = {}^5/_2$ (*260*) having nine F2*p*-like MO calculated to have I between 16.6 and 18.3 eV, the five (half occupied) Fe3*d*-like orbitals 26.6, 28.2 and 28.3 eV, followed by three F2*s*-like MO at 41.6 and 41.7 eV. This picture is almost a caricature made by an atomic spectroscopist ($I_3$ of gaseous $Fe^{+2}$ is 30.65 and $I_4$ of $Fe^{+3}$ is 54.8 eV) and brings into the mind the discussion by *Dunn* (*261*) of the origin of partly covalent bonding in iron(III) compounds. However, it is remember-ed that the calculated $I$-values (*260*) are *Koopmans* values assuming "frozen" orbitals. The $I$(Fe3*d*) calculated for adapting orbitals, including the rearrangement energy, undoubtedly would be closer to 18 eV. Another conclusion of this work is that the modification of the fractional atomic charges by covalent bonding can change $I$ to a larger extent than bonding or anti-bonding effects. This is the major reason for $I$(3*d*) suddenly becoming much larger (Table 3) for Zn(II) and Ga(III) than for Cu(II) and Co(III), creating large difficulties (*30*) for the concept of "ligand field stabilization" using the closed-shell $3d^{10}$ as standard of compari-son, and justifying some of the critical remarks by *Williams* (*262*). Further on, the polyatomic entities present the same feature as known from *Hartree-Fock* calculations for atoms (*1*) that it is the total energy of the whole wave-function which is minimized, and not the individed orbitals, which may very well increase their energy in a few cases during the minimization process.

Besides the large difference $A_*(d,d)$ or $A_*(f,f)$ between $I$ and the electron affinity of a partly filled shell, it is clear that there is still another characteristic conserving the individuality of the *d* and *f* orbitals in compounds. This may very well be the large angular kinetic energy from Eq. (12) equal to three (*d*) or six (*f*) times $\langle r^{-2}\rangle$ in atomic units. Dr. *Richard Watson* was as kind, for the purpose of comparison (*118*) with photo-electron signal intensities, as to provide $\langle r^{-2}\rangle$ for various *Hartree-Fock* orbitals. Table 4 gives these results in eV showing that $E_{\text{ang kin}}$ is 1.4 to 3 times larger than $I(2p)$ going backwards from arsenic to sodium, whereas this ratio is 25 for $I$(3*d*) in $Fe^{+2}$ and $Cu^+$ decreasing toward 3.3 in $Cd^{+2}$. By the same token, $E_{\text{ang kin}}$ is more than 10 times $I$(4*d*) in $Ag^+$ and $Cd^{+2}$ and still 4 times $I$(4*d*) in $Gd^{+3}$. The latter ion provides our only example of the angular kinetic energy being almost 40 times larger than $I$(4*f*). Said in other words, the partly filled shell has the kinetic energy [of which the angular part is a proportion, *cf.* Eq. (15)] far larger than the ionization energy. Since the virial theorem Eq. (3) is valid for the total wave-function, one might have expected the kinetic energy of the individual electron to be (*4*) half the expression:

Table 4. *Hartree-Fock* average $\langle r^{-2}\rangle$ in bohr$^{-2}$ according to *Watson* (*118*), the angular kinetic energy and the ionization energy (*15*) in eV, the number $(2j+1)$ or $(4l+2) = n_e$ of electrons giving photo-electron signals with intensities relative to fluorine 1*s* using photons originating in an aluminium (*118*) or magnesium (*119*) anti-cathode

| Species | Shell | $\langle r^{-2}\rangle$ | $E_{\text{kin ang}}$ | $I(nl)$ | $n_e$ Electrons | Intensity 1486.6 eV | 1253.6 eV | 100 · intensity 1486.6 eV | $n_e\langle r^{-2}\rangle$ 1253.6 eV |
|---|---|---|---|---|---|---|---|---|---|
| $Na^+$ | 2*p* | 4.2 | 117 | 36 | 6 | 0.08 | 0.14 | 0.3 | 0.5 |
| $Cl^0$ | 2*p* | 14.3 | 390 | 203 | (6) | 0.42 | 0.44 | 0.5 | 0.5 |
| | 3*p* | 1.19 | 32 | 10 | 6 | 0.05 | 0.04 | 0.7 | 0.6 |
| $Fe^{+2}$ | 2*p* | 40.3 | 1110 | 716 | 4 | 1.8 | 1.15 | 1.1 | 0.7 |
| | 3*p* | 5.4 | 147 | 62 | 6 | 0.25 | 0.32 | 0.8 | 1.0 |
| | 3*d* | 2.13 | 175 | 7 | 6 | 0.06 | 0.08 | 0.5 | 0.6 |
| $Cu^+$ | 2*p* | 51.5 | 1400 | 938 | 4 | 3 | 1.9 | 1.5 | 0.9 |
| | 3*p* | 7.2 | 200 | 81 | (6) | 0.4 | 0.6 | 0.9 | 1.4 |
| | 3*d* | 2.72 | 225 | 9 | 10 | 0.15 | 0.4 | 0.5 | 1.3 |
| $As^0$ | 2*p* | 69.3 | 1900 | 1330 | 4 | 3 | — | 1.1 | — |
| | 3*p* | 10.6 | 290 | 150 | 4 | 0.38 | 0.33 | 0.9 | 0.8 |
| | 3*d* | 4.8 | 390 | 51 | 10 | 0.48 | 0.45 | 1.0 | 0.9 |
| $Br^-$ | 3*p* | 12.6 | 340 | 187 | 4 | 0.5 | 0.5 | 1.0 | 1.0 |
| | 3*d* | 6.1 | 500 | 74 | 10 | 0.55 | 0.6 | 0.9 | 1.0 |
| | 4*p* | 0.91 | 25 | 9 | 6 | 0.08 | 0.11 | 1.4 | 2.0 |
| $Rb^+$ | 3*p* | 14.9 | 405 | 243 | 4 | 0.7 | 0.6 | 1.1 | 1.0 |
| | 3*d* | 7.4 | 600 | 116 | 6 | 0.6 | 0.66 | 1.3 | 1.5 |
| | 4*p* | 1.59 | 43 | 18 | 6 | 0.1 | 0.18 | 1.0 | 1.8 |
| $Ag^+$ | 3*p* | 28.9 | 800 | 578 | 4 | 0.8 | 0.4 | 0.7 | 0.4 |
| | 3*d* | 15.6 | 1280 | 373 | 6 | 2.5 | 2.45 | 2.7 | 2.6 |
| | 4*p* | 4.9 | 133 | 66 | (6) | 0.1 | 0.09 | 0.3 | 0.3 |
| | 4*d* | 1.83 | 150 | 10 | 10 | 0.35 | 0.3 | 1.9 | 1.6 |
| $Cd^{+2}$ | 3*p* | 30.6 | 830 | 623 | 4 | 1.0 | 0.6 | 0.8 | 0.5 |
| | 3*d* | 16.6 | 1360 | 411 | 6 | 3 | 3.35 | 3.0 | 3.4 |
| | 4*p* | 5.4 | 147 | 74 | (6) | 0.1 | 0.1 | 0.3 | 0.3 |
| | 4*d* | 2.14 | 175 | 17 | 10 | 0.5 | 0.6 | 2.3 | 2.8 |
| $Gd^{+3}$ | 3*d* | 36.1 | 2950 | 1193 | 6 | 3 | — | 1.4 | — |
| | 4*p* | 14.8 | 400 | 275 | 4 | 0.4 | — | 0.7 | — |
| | 4*d* | 7.5 | 610 | 149 | (6) | 0.8 | 1.0 | 1.8 | 2.2 |
| | 5*p* | 2.2 | 60 | 28 | (6) | 0.1 | 0.24 | 0.8 | 1.8 |
| | 4*f* | 3.3 | 540 | 14 | 7 | 0.3 | 0.48 | 1.3 | 2.0 |

$$H(a) = Z\langle r^{-1}\rangle - \frac{1}{2}\sum_{k \neq a} A_*(k,a) \tag{29}$$

which is the arithmetic average of the one-shot ionization energy $I(a)$ from Eq. (7) and the difference $[Z\langle r^{-1}\rangle - E_{\text{kin}}(a)]$ between the nuclear attraction and the kinetic energy of the electron considered. However, this is not the whole truth, though the sum of $H(a)$ for all $q$ electrons in a monatomic entity equals twice the total kinetic energy, and hence also twice the sum of all the consecutive ionization energies. For instance, the excited configuration [Ne]4$f$ of the sodium atom has very closely the $I(4f)$ value (1/32) hartree characterizing the hydrogen atom, whereas $^1/_2$ H(4$f$) from Eq. (29) is approximately (11 – 5)/32 hartreee, about 6 times larger. Rather, the coupled integro-differential equations of *Hartree* and *Fock* (*1*) have such a structure that the repulsion between an interior and an exterior position of two electrons is not distributed equally, as in the quantity $H(a)$ of Eq. (29), but rather that the outer electron sees a central field $U(r)$ as if the whole $A_*(k,a)$ was substracted from $Z\langle r^{-1}\rangle$.

The situation is even more extreme in the 4$f$ group. With exception of $4f^1$ Ce(III), $4f^2$Pr(III) and $4f^8$Tb(III), the $I$ of the loosest bound ligand orbitals (such as 2$p$ of oxide or fluoride) are *lower* than $I(4f)$ as independently found at the Bell laboratories (*263*) and in Geneva (*15, 31*). Since it is known from the weak nephelauxetic effect in $4f^q$ transitions (*4, 21*) and in particular the spin-orbit coupling varying less than one percent in ytterbium(III) complexes (*136*) that the combined effect of delocalization of the anti-bonding 4$f$ orbitals and the expanded radial function (corresponding to decrease of the fractional atomic charge) is rather insignificant. Though it may be argued that $S_{MX}$ in Eq. (28) is small, it is rather serious that the ionized system lacking a 4$f$ electron corresponds to the same eigen-value as other states lacking a ligand electron, since it is a property of the *Schrödinger* equation that an arbitrary linear combination of degenerate eigen-states also is a solution. Hence, it is very difficult to find an objective method (*30*) to determine the eigen-vectors in the LCAO model for 4$f$ group compounds.

A comparable problem occurs in the metallic lanthanides (*71*) which all have positive $I^*(4f)$ values relative to the *Fermi* level, though they are moderate in metallic europium and ytterbium exemplifying the conditional oxidation states (*21*) $4f^7$Eu[II] and $4f^{14}$Yb[II], in contrast to the other M[III]. The physical reason why the partly filled 4$f$ shell does not get invaded by the conduction electrons with lower $I$ is that the electron affinity (roughly $I(4f) - A_*(4f,4f)$ with corrections (*2, 142*) for inter-atomic relaxation effects) of the partly filled shell is smaller than the work function, or perhaps even negative. One can compare $I'(4f)$ values for oxides and fluorides and $I^*(4f)$ for the metallic elements (*71*). It is discussed in another volume of this series (*138*) how the final states of $4f^{q-1}$ obtained by ionization are selected from the many alternatives of Eq. (22) according to the square of the coefficients of the fractional parentage, as also discussed by *Cox* in this volume. It is important to note (*71*) that the nephelauxetic effect is more important in the final states of ionization $4f^{11}$Tm[IV] and $4f^{12}$Yb[IV] than in the cases of $4f^7$Tb[IV] and $4f^8$Dy[IV] corresponding to oxidation states also known from chemical compounds.

The characteristic variation of the ionization energy of $4f^q$ as a function of $q$ can be explained (*28, 143*) by the *refined spin-pairing energy theory* obtained from Eq. (21) by adding further parameters describing the distance from the barycentre of the maximum $S$ value to the lowest $J$-level of the lowest term. *Nugent, Baybarz, Burnett* and *Ryan* (*264*) have calculated $I$-values from the refined spin-pairing energy theory partly derived from uncorrected optical elec-tronegativities as defined in Eq. (23). When inserting $x_{uncorr}$(M) in Eq. (24), one has to consider the electron transfer spectra of M(III) when discussing $I$(M4$f$) of M(II), and of M(IV) when discussing $I$ of M(III).

It is possible to define a *chemical ionization energy* (*265*)

$$I_{\text{chem}} = E^0 + 4.5 \text{ eV} \tag{30}$$

by adding a constant 4.5 eV (known within a-tenth eV) representing the actual ionization energy of the standard hydrogen electrode, to $E^0$, the standard oxidation potential of $Fe(H_2O)_6^{+2}$, Eu(II) aqua ions, and other species such as $RuO_4^-$ or $IrCl_6^{-3}$. In the case of transition group aqua ions, it turns out (*21, 265*) that $I_{\text{chem}}$ going from the oxidation state $(z-1)$ to $z$ can be written to a good approximation $(2z-1)\varkappa$ with the parameter $\varkappa$ dependent on the group:

$$\varkappa = 5.3 \text{ eV } (3d) \quad 4.3 \text{ eV } (4f) \quad 3.8 \text{ eV } (5f) \tag{31}$$

as if the hydration energy of gaseous $M^{+z}$ is $-\varkappa z^2$ as if the differential variation of ionic radii with $z$ can be neglected in a dielectric description. This concept can be developed in the theory of *hydration energy differences* which is also connected with *Ahrland's softness parameters* providing a quantitative measure of *Pearson's* classification (*76, 77*) based exclusively on hydration energies and ionization energies $I_z$ of gaseous ions.

Seen from the point of view of photo-electron spectra, $I_{\text{chem}}$ is an adiabatic ionization energy allowing the internuclear distances to re-arrange. Hence, it is not surprising that the "vertical" $I$ may be anywhere from 1 to 5 eV above the corresponding $I_{\text{chem}}$. A relatively clear-cut case with the minimum of hydration energy of the large, univalent cation is $Fe(C_5H_5)_2^+$ formed in aqueous solution with $E^0 = 0.40$ V and hence $I_{\text{chem}} = 4.9$ eV to be compared with $I = 6.88$ eV for gaseous $Fe(C_5H_5)_2$ in Table 3. The width of this signal (*227*) suggest an adiabatic $I$ somewhere close to 6.6 eV or a hydration energy of the cation 1.7 eV. For comparison, it may be noted that the free energy of hydration of $I^-$ is 2.7 eV to be compared with 3.3 eV for a perfect dielectric receiving an ion with the radius 2.20 Å, whereas the $Cs^+$ with $r_{\text{ion}} = 1.65$ Å would be 4.4 eV in a perfect dielectric, but as matter of fact has the lower free energy of hydration 3.0 eV. The heat of hydration of a gaseous proton is 11.3 eV, and of $Al^{+3}$ 48.3 eV (*79*) corresponding to a much stronger interaction with the solvent (*30*). It is not surprising if $Fe(C_5H_5)_2^+$ with an effective radius about 3 Å shows a hydration energy 1.7 eV, in particular because the dielectric stabilization is partly compensated by the hydrophobic cation disturbing the adjacent water molecules.

## C. Energy Bands and Relevant Symmetry

Solid-state physicists consider *Bloch's* energy bands the logical extension of MO theory to large crystals. Though approximations such as the "augmented plane-wave method" (where the wave-functions of free electrons are orthogonalized on the core orbitals of the constituent atoms) present the qualification of being almost the only serious attempt of directly solving *Schrödinger's* equation before the $X\alpha$ method of *Slater* and *Johnson* (*112, 113*), the intrinsic difficulty is that MO in small molecules serve to classify excitations and ionizations on the explicit condition that the symmetry is chosen in a way avoiding unneccessary correlation effects (*4, 7*). A classical case is a hydrogen molecule stretched to a large internuclear distance $R$, where the total wave-function is a mixture of, say, 51 percent $(\sigma_g)^2$ and 49 percent $(\sigma_u)^2$ and is *better* described as two hydrogen atoms with each a 1$s$ electron. The text-book argument that a closed-shell anti-symmetrized *Slater* determinant is invariant for a unitary transformation forming linear combinations of the orbitals has the fundamental weak point that the actual total $\Psi$ *is not* a *Slater* determinant. The question of the nuclear equilibrium positions in a molecule not necessarily representing the relevant point-group has a variety of aspects (*7*), it is very difficult to prevent chemists from saying that a given molecule *almost* has a high symmetry, as is always true when including vibrations.

In the case of $d$ and $f$ shells, the mechanism of obtaining an integer $q$ of electrons is the coefficient $q(q-1)/2$ in Eq.(5) paying a premium for highly localized electrons with large $\langle r^{-1}\rangle$. The transition (as a function of increasing internuclear distances of decreasing temperature) from a metallic to an insulator with weak "hopping" semiconductivity (described by *Mott*) is not unrelated to the competition between the disorder corresponding to the *Boltzmann* population of states of various energy and the tendency (gaining at low temperature) to exemplify the lowest state. Though the groundstate degeneracy number $k$ corresponds to $N^k$ state of $N$ atoms containing the partly filled shell, this is a very minute proportion (*7*) of the number of states in an energy band description.

Photo-electron spectra have not been particularly favourable to energy band pictures, and suggest the attitude of a mathematics teacher reminding that before calculating the logarithm of some odd quantity, one must ask the question whether the logarithm exists (or subsists, as Medieval philosophers would say). However, there is one clear-cut effect of energy bands in the 5$d$ (valence) region of gold (*66*) that the two maxima are separated about twice as much as the spin-orbit coupling would account for, and that this distance is attenuated as a function of decreasing gold concentration in alloys, as also discussed by *Watson* and *Perlman* in this volume. In most other cases, the measurements support the idea behind the title "Persistence of Atomic Orbitals in Complexes and Other Compounds" of a recent survey (*266*).

## D. Description of Chemical Bonding vs. One-electron Ionization

Photo-electron spectrometry and chemistry look at the same compounds with rather different eyes. This is not only because the heat of dissociation to atoms (excepting $H_2$) is negligible compared with the energy (roughly $-Z^{2.4}$ rydberg) of an atom compared with its separated electrons and nucleus. It is also because quantum mechanics treats "black box gambling machines" having definite probabilities of various outcomes of putting in a slug. It is remembered that the typical soft X-ray photon traverses about thousand atoms before getting absorbed in a definite atom. This quantum jump then produces the observable $E_{\text{kin}}$ of the ejected electron. Seen from the point of view of the small child pulling an alarm clock to pieces, one may ask the rather pernicious question whether the orbital was there before the ionization took place. However, before getting too sceptical, one should realize that quantum mechanics provides related paradoxes (*18*). For instance, a heteropolar diatomic molecule such as HCl has no observable electric dipole moment in the lowest rotational state, though most chemists persist in believing in a "metaphysical" dipole moment being conserved inside the spherically symmetric, rotational wave-function, However, a much more realistic argument is that electron configurations classify in all details the numerous discrete energy levels of a monatomic entity (*26*).

A compromise between considering orbitals as hard-ware components of a molecule or a crystal and considering them as epiphenomena produced by the photon may be found in the book (*267*) by *Linderberg* and *Öhrn* discussing the singularities of the excitability as a function of the excitation energy of many-electron systems. Such a treatment can go beyond *Hartree-Fock* accuracy having slightly lower energies of the singularities, corresponding to the relaxation effects decreasing $I$ below the *Koopmans* value. It is particularly interesting that one may run into singularities corresponding to less than one electron, due to accidental coincidences of different ionized configurations. In the specific case of the xenon atom, *Gelius* (*268*) believes that the straightforward ionization to [Ar]$3d^{10}4s^2 4p^5$ $4d^{10}5s^2 5p^6$ removing a $4p$ electron shows satellites due to the adjacent configuration [Ar]$3d^{10}4s^2 4p^6 4d^8 5s^2 5p^6 4f$. It has also been argued (*174*) that the $3s$ region of manganese(II) shows additional signals due to configuration mixing with [Ne]$3s^2 3p^4 3d^6$ which is a kind of intra-atomic electron transfer process.

With exception of the somewhat controversial satellite structures, photo-electron spectra induced by high-energy photons do not provide information about empty orbitals. This is a major simplification relative to spectroscopy in the visible, ultra-violet and X-ray regions, but at the same time, it contributes to the chemist's feeling that the picture obtained is somewhat incomplete. On the other hand, the evidence for the verifiable character of penultimate MO is overwhelmingly massive, and the valence region is now much better understood.

There is a last aspect where photo-electron spectra furnish unexpected information. That is the very short time-scale, where *Campagna et al.* (*269*) have found that the metallic, stoichiometric compound TmTe (crystallizing in the cubic NaCl type) on an instantaneous picture contains about equal amounts of $4f^{13}$ Tm[II] and of $4f^{12}$Tm[III], the latter providing the conduction electrons. The "hopping" of an electron from (the crystallographically) equivalent $4f^{13}$ to $4f^{12}$ sites must be very rapid. Perhaps the most astonishing property (*136, 138*) of this unusual compound is that the nephelauxetic effect is only marginally (say by one percent) larger on the Tm[III] formed by photo-ionizing Tm[II] of TmTe than known from the absorption spectra (*4*) of thulium(III) compounds, whereas the nephelauxetic ratio seems to be 0.90 for the Tm[IV] formed from Tm[III] in TmTe, almost the same as 0.87 for metallic thulium (*71, 138*). The different nephelauxetic behaviour of the two kinds of thulium in the telluride is a magnificent confirmation of the *Copenhagen* principle of final states.

It has been fancied (*28*) that the strongly magnetic $d$-group metals such as iron, cobalt and nickel have asymmetric photo-electron signals both of inner shells and in the valence shell (*237, 238*) because one sees adjacent signals of $3d^6$Fe[II], $3d^7$Fe[I] and $3d^8$Fe[0] co-existing on an instantaneous picture like the TmTe situation. However, it is probable that the MO (energy band) description makes the atoms of these elements really equivalent, like the two mercury atoms in $Hg_2^{+2}$. Nevertheless, this type of question needs comparison with experimental evidence almost every time, the cases being *sui generi*. Cluster compounds such as $Nb_6Cl_{12}^{+2}$ and $Nb_6Cl_{12}^{+3}$ kindly forwarded to us by Dr. *A. J. Thomson* at the University of East Anglia, Norwich clearly show non-equivalent Nb signals, as if an instantaneous picture contained M[II] and M[III]. However, in this particular case (like in the non-equivalent terminal and central nitrogen atoms of NNO and $N_3^-$) one cannot be perfectly certain that one does not see an effect of differing fractional charges rather than well-defined $d^3$ and $d^2$ configurations.

The complicated consequences of the absence of an electron in an inner shell are well-known from atomic line spectra. Thus, one may consider the red light emitted by neon atoms a perturbation of the two closely adjacent transitions in the yellow due to [Ne]$3p \rightarrow$ [Ne]$3s$ in sodium atoms, by removing a proton from the nucleus, and a $2p$ electron. The excited configuration $1s^2 2s^2 2p^5 3p$ containing 10 $J$-levels has 30 allowed transitions to the four $J$-levels of the less excited configuration $1s^2 2s^2 2p^5 3s$ of the neon atom (*25, 26*). The wide distribution (*134, 137*) of $J$-levels of $4d^9 4f^q$ formed by $4d$ ionization of lanthanides is a large-scale example of such (interelectronic repulsion and spin-orbit) coupling of a vacancy with another shell. It is indeed fascinating that chemically very stable gadolinium(III) compounds have a half-filled $4f$ shell with $\langle r^{-1} \rangle = 1.6$ bohr$^{-1}$ as if they had been hit by X-rays. However, the $4f$ orbitals have a very small electron affinity because of large $A_*(4f,4f)$ and exceedingly large angular kinetic energy.

*Acknowledgments.* The Swiss National Science Foundation provided the grant for the Varian IEE-15 photo-electron spectrometer installed in Geneva 1971. The writer is grateful to Dr. *Hervé Berthou* for careful experimental work and for many discussions about *Auger* transitions.

## V. References

1. *Hartree, R. D.:* The calculation of atomic structures. New York: John Wiley 1957.
2. *Jørgensen, C. K.:* Advan. Quantum Chem. *8*, 137 (1974).
3. *Koopmans, T.:* Physica *1*, 104 (1933).
4. *Jørgensen, C. K.:* Orbitals in atoms and molecules. London: Academic Press 1962.
5. *Herigonte, P., v.:* Struct. Bonding *12*, 1 (1972).
6. *Turner, D. W., Baker, C., Baker, A. D., Brundle, C. R.:* Molecular photoelectron spectroscopy. London: Interscience 1970.
7. *Jørgensen, C. K.:* Modern aspects of ligand field theory. Amsterdam: North-Holland Publishing Co. 1971.
8. *Herzberg G.:* Electronic spectra and electronic-structure of polyatomic molecules. Princeton: Van Nostrand 1966.
9. *Lindholm, E., Fridh, C., Åsbrink, L.:* Discussions Faraday Soc. *54*, 127 (1973).
10. *Potts, A. W., Price, W. C., Streets, D. G., Williams, T. A.:* Discussions Faraday Soc. *54*, 168 (1973).
11. *Jarvie, J., Wilson, W., Doolittle, J., Edminston, C.:* J. Chem. Phys. *59*, 3020 (1973).
12. *Siegbahn, K., Nordling, C., Fahlman, A., Nordberg, R., Hamrin, K., Hedman, J., Johansson, G., Bergmark, T., Karlsson, S. E., Lindgren, I., Lindgren, B.:* ESCA-atomic, molecular and solid-state structures studied by means of electron spectroscopy. Stockholm: Almqvist & Wiksell 1967.

13. *Jørgensen, C. K.:* Absorption spectra and chemical bonding in complexes. Oxford: Pergamon Press 1962.
14. *Jørgensen, C. K.:* Advan. Chem. Phys. *5*, 33 (1963).
15. *Jørgensen, C. K., Berthou, H.:* Mat. Fys. Medd. Danske Vidensk. Selskab *38*, no. 15 (1972).
16. *Manne, R., Åberg, T.:* Chem. Phys. Letters *7*, 282 (1970).
17. *Ballentine, L. E.:* Foundations Phys. *3*, 229 (1973).
18. *Jørgensen, C. K.:* Theoret. Chim. Acta *34*, 189 (1974).
19. *Greenwood, N. N., Gibb, T. C.:* Mössbauer spectroscopy. London: Chapman & Hall 1971.
20. *Burger, K.:* Coordination chemistry-experimental Methods. Cleveland: Chemical Rubber Co. Press 1973.
21. *Jørgensen, C. K.:* Oxidation numbers and Oxidation states. Berlin-Heidelberg-New York: Springer 1969.
22. *Jørgensen, C. K.:* Angew. Chem. *85*, 1 (1973); Angew. Chem. Intern. Ed. Engl. *12*, 12 (1973).
23. *Politzer, P., Daiker, K. C.:* Chem. Phys. Letters *20*, 309 (1973).
24. *Ruedenberg, K.:* Rev. Mod. Phys. *34*, 326 (1962).
25. *Condon, E. U., Shortley, G. H.:* Theory of atomic spectra (2. ed.). Cambridge: University Press 1953.
26. *Moore, C. E.:* Atomic energy levels. Natl. Bur. Std. (U.S.), Circ. *467*, volumes 1,2,3. Washington, D. C.: 1949, 1952, 1958.
27. *Moore, C. E.:* Ionization potentials and ionization limits derived from the analyses of optical spectra. NSRDS—NBS 34. Washington, D. C.: Natl. Bur. Std. 1970.
28. *Jørgensen, C. K.:* Struct. Bonding *13*, 199 (1973).
29. *Stoner, E. C.:* Phil. Mag. *48*, 719 (1924).
30. *Jørgensen, C. K.:* Chimia (Aarau) *28*, 6 (1974).
31. *Jørgensen, C. K.:* Chimia (Aarau) *27*, 203 (1973).
32. *Gillespie, R. J.:* Molecular geometry. London: Van Nostrand 1972.
33. *Jørgensen, C. K.:* Progr. Inorg. Chem. *4*, 73 (1962).
34. *Jørgensen, C. K.:* Helv. Chim. Acta Fasc. extraord. *Alfred Werner* 131 (1967).
35. *Finn, P., Pearson, R. K., Hollander, J. M., Jolly, W. L.:* Inorg. Chem. *10*, 378 (1971).
36. *Allison, D. A., Johansson, G., Allan, C. J., Gelius, U., Siegbahn, H., Allison, J., Siegbahn, K.:* J. Electron Spectry. *1*, 269 (1973).
37. *Chadwick, D.:* Chem. Phys. Letters *21*, 291 (1973).
38. *Kramer, L. N., Klein, M. P.:* J. Chem. Phys. *51*, 3618 (1969).
39. *Fadley, C. S., Hagström, S. B. M., Klein, M. P., Shirley, D. A.:* J. Chem. Phys. *48*, 3779 (1968).
40. *Folkesson, B.:* Acta Chem. Scand. *27*, 287 (1973).
41. *Stucky, G. D., Matthews, D. A., Hedman, J., Klasson, M., Nordling, C.:* J. Am. Chem. Soc. *94*, 8009 (1972).
42. *Rabinowitch, E., Thilo, E.:* Periodisches System, Geschichte und Theorie. Stuttgart: Ferdinand Enke 1930.
43. *Sugar, J., Reader, J.:* J. Chem. Phys. *59*, 2083 (1973).
44. *Bethe, H.:* Ann. Physik *3*, 133 (1929).
45. *Gerloch, M., Slade, R. C.:* Ligand-field parameters. Cambridge: University Press 1973.
46. *Berry, R. S., Reimann, C. W., Spokes, G. N.:* J. Chem. Phys. *37*, 2278 (1962).
47. *Milstein, R., Berry, R. S.:* J. Chem. Phys. *55*, 4146 (1971).
48. *Jørgensen, C. K.:* Chimia (Aarau) *25*, 213 (1971).
49. *Citrin, P. H., Thomas, T. D.:* J. Chem. Phys. *57*, 4446 (1972).
50. *Nefedov, V. I.:* Application of the X-ray photoelectron spectroscopy in chemistry. In: Structure of molecules and the chemical bond Moscow: VINITI Editions 1973.
51. *Poole, R. T., Jenkin, J. G., Leckey, R. C. G., Liesegang, J.:* Chem. Phys. Letters *22*, 101 (1973).
52. *Poole, R. T., Liesegang, J., Leckey, R. C. G., Jenkin, J. G.:* Chem. Phys. Letters *23*, 194 (1973).
53. *Mott, N. F., Gurney, R. W.:* Electronic processes in ionic crystals (2. ed.). Oxford: Clarendon Press 1948.

54. *Butler, M. A., Wertheim, G. K., Rousseau, D. L., Hüfner, S.:* Chem. Phys. Letters *13*, 473 (1972).
55. *Hnatowich, D. J., Hudis, J., Perlman, M. L., Ragaini, R. C.:* J. Appl. Phys. *42*, 4883 (1971).
56. *Kim, K. S., O'Leary, T. J., Winograd, N.:* Anal. Chem. *45*, 2214 (1973).
57. *Betteridge, D., Carver, J. C., Hercules, D. M.:* J. Electron Spectry. *2*, 327 (1973).
58. *Siegbahn, K., Nordling, C., Johansson, G., Hedman, J., Hedén, P. F., Hamrin, K., Gelius, U., Bergmark, T., Werme, L. O., Manne, R., Baer, Y.:* ESCA Applied to free molecules. Amsterdam: North-Holland Publishing Co. 1969.
59. *Thomas, T. D.:* J. Chem. Phys. *52*, 1373 (1970).
60. *Bremser, W., Linnemann, F.:* Chemiker-Zg. *95*, 1011 (1971).
61. *Tovborg Jensen, A.:* Arkiv Kemi *30*, 165 (1968).
62. *Johansson, G., Hedman, J., Berndtsson, A., Klasson, M., Nilsson, R.:* J. Electron Spectry. *2*, 295 (1973).
63. *Evans, S.:* Chem. Phys. Letters *23*, 134 (1973).
64. *Jørgensen, C. K., Berthou, H.:* Chem. Phys. Letters *31*, 416 (1975).
65. *Citrin, P. H.:* Phys. Rev. Letters *31*, 1164 (1973).
66. *Berthou, H., Jørgensen, C. K.:* J. Electron Spectry. *5*, 935 (1974).
67. *Schwartz, M. E.:* Chem. Phys. Letters *5*, 50 and *7*, 78 (1970).
68. *Shirley, D. A.:* Chem. Phys. Letters *16*, 220 (1972).
69. *Aarons, L. J., Guest, M. F., Hall, M. B., Hillier, I. M.:* J. C. S. Trans. Faraday Soc. II *69*, 563 (1973).
70. *Fadley, C. S., Shirley, D. A.:* Phys. Rev. *A 2*, 1109 (1970).
71. *Cox, P. A., Baer, Y., Jørgensen, C. K.:* Chem. Phys. Letters *22*, 433 (1973).
72. *Brodén, G., Hagström, S. B. M., Norris, C.:* Physik kondens. Materie *15*, 327 (1973).
73. *Citrin, H., Hamann, D. R.:* Chem. Phys. Letters *22*, 301 (1973).
74. *Citrin, H., Hamann, D. R.:* Phys. Rev. B *10*, 4948 (1974).
75. *Jørgensen, C. K., Berthou, H., Balsenc, L.:* J. Fluorine Chem. *1*, 327 (1972).
76. *Pearson, R. G.:* J. Am. Chem. Soc. *85*, 3533 (1963).
77. *Pearson, R. G.:* Hard and soft acids and bases (Benchmark papers in Inorganic Chemistry). Stroudsburg: Dowden, Hutchinson & Ross 1973.
78. *Salzmann, J. J., Jørgensen, C. K.:* Helv. Chim. Acta *51*, 1276 (1968).
79. *Jørgensen, C. K.:* Topics Curr. Chem. *56*, 1 (1975).
80. *Jørgensen, C. K., Judd, B. R.:* Mol. Phys. *8*, 281 (1964).
81. *Peacock, R. D.:* Struct. Bonding *22*, 81 (1975).
82. *Jørgensen, C. K., Berthou, H.:* J. Fluorine Chem. *2*, 425 (1973).
83. *Cauchois, Y.:* J. Phys. Radium *13*, 113 (1952).
84. *Cauchois, Y.:* J. Phys. Radium *16*, 253 (1956).
85. *Bearden, J. A., Burr, A. F.:* Rev. Mod. Phys. *39*, 125 (1967).
86. *Werme, L. O., Grennberg, B., Nordgren, J., Nordling, C., Siegbahn, K.:* Nature *242*, 453 (1973).
87. *Werme, L. O., Grennberg, B., Nordgren, J., Nordling, C., Siegbahn, K.:* Phys. Rev. Letters *30*, 523 (1973).
88. *Fuggle, J. C., Burr, A. F., Watson, L. M., Fabian, D. J., Lang, W.:* J. Phys. *F 4*, 335 (1974).
89. *Bonnelle, C.:* Ann. Phys. (Paris) *1*, 439 (1966).
90. *Hayes, W., Brown, F. C., Kunz, A. B.:* Phys. Rev. Letters *27*, 774 (1971).
91. *Watson, R. E., Perlman, M. L.:* Research applications of Synchrotron radiation. BNL 50381. Springfield (Virginia): National Technical Information Service 1973.
92. *Sagawa, T., Kato, R., Sato, S., Watanabe, M., Ishii, T., Nagakura, I., Kono, S., Suzuki, S.:* J. Electron Spectry. (Namur 1974 Supplement) *5*, 551 (1974).
93. *Satoko, C., Sugano, S.:* J. Phys. Soc. Japan *34*, 701 (1973).
94. *Richard, P.:* Phys. Fennica *9(S1)*, 3 (1974).
95. *Ziem, P., Stolterfoht, N., Ridder, D.:* Phys. Fennica *9(S1)*, 39 (1974).
96. *Briand, J. P., Chevallier, P., Johnson, A., Rozet, J. P., Tavernier, M., Touati, A.:* Phys. Fennica *9(S1)*, 409 (1974).
97. *Faessler, A.:* Angew. Chem. *84*, 51 (1972).

98. *Stelling, O.:* Z. Physik *50*, 506 (1928).
99. *Hagström, S., Nordling, C., Siegbahn, K.:* Z. Physik *178*, 439 (1964).
100. *Sugar, J.:* Phys. Rev. *A 6*, 1764 (1972).
101. *Bonnelle, C., Karnatak, R. C., Sugar, J.:* Phys. Rev. *A 9*, 1920 (1974).
102. *Sugar, J.:* Phys. Rev. *B 5*, 1785 (1972).
103. *Böke, K.:* Z. Physik. Chem. (Frankfurt) *10*, 45 and 59; *11* 326 (1957).
104. *Fiedler, G.:* Z. Physik. Chem. (Frankfurt) *37*, 79 (1963).
105. *Jørgensen, C. K.:* Progr. Inorg. Chem. *12*, 101 (1970).
106. *Best, P. E.:* J. Chem. Phys. *44*, 3248 (1966).
107. *Cauchois, Y., Bonnelle, C.:* Colloques Internationaux CNRS (Paris) *191*, 125 (1970).
108. *Rauk, A., Allen, L. C., Clementi, E.:* J. Chem. Phys. *52*, 4133 (1970).
109. *Cauchois, Y.:* J. Chim. Phys. *51*, D 76 (1954); J. Physique Radium *13*, S 73 (1952).
110. *LaVilla, R. E., Deslattes, R. D.:* J. Chem. Phys. *44*, 4399 (1966).
111. *Sachenko, V. P., Polozhentsev, E. V., Kovtun, A. P., Migal, Yu. F., Kolesnikov, V. V., Vedrinski, R. V.:* Phys. Fennica *9(S1)*, 129 (1974).
112. *Slater, J. C.:* Advan. Quantum Chem. *6*, 1 (1972).
113. *Johnson, K. H.:* Advan. Quantum Chem. *7*, 143 (1973).
114. *Dehmer, J. L.:* Phys. Fennica *9(S1)*, 60 (1974).
115. *Mayer, M. G.:* Phys. Rev. *60*, 184 (1941).
116. *Teegarden, K., Baldini, G.:* Phys. Rev. *155*, 896 (1967).
117. *Wagner, C. D.:* Anal. Chem. *44*, 1050 (1972).
118. *Jørgensen, C. K., Berthou, H.:* Discussions Faraday Soc. *54*, 269 (1973).
119. *Berthou, H., Jørgensen, C. K.:* Anal. Chem. *47*, 482 (1975).
120. *Manson, S. T., Cooper, J. W.:* Phys. Rev. *165*, 126 (1968).
121. *Fano, U., Cooper, J. W.:* Rev. Mod. Phys. *40*, 441 (1968).
122. *Krishna, V. G.:* J. Chem. Phys. *50*, 792 (1969).
123. *Murrell, J. N., Ralston, B. J.:* J. C. S. Trans. Faraday Soc. II *68*, 1393 (1972).
124. *Jørgensen, C. K.:* Theoret. Chim. Acta *24*, 241 (1972).
125. *Davis, D. W., Shirley, D. A.:* J. Chem. Phys. *56*, 669 (1972).
126. *Verhaegen, C.:* J. Chem. Phys. *49*, 4696 (1968).
127. *Racah, G.:* Phys. Rev. *76*, 1352 (1949).
128. *Fadley, C. S., Shirley, D. A., Freeman, A. J., Bagus, P. S., Mallow, J. V.:* Phys. Rev. Letters *23*, 1397 (1969).
129. *Carver, J. C., Schweitzer, G. K., Carlson, T. A.:* J. Chem. Phys. *57*, 973 (1972).
130. *Helmer, J. C.:* J. Electron Spectry. *1*, 259 (1973).
131. *Watson, R. E.:* Phys. Rev. *118*, 1036 (1960).
132. *Cohen, R. L., Wertheim, G. K., Rosencwaig, A., Guggenheim, H. J.:* Phys. Rev. *B 5*, 1037 (1972).
133. *Bonnelle, C., Jørgensen, C. K.:* J. Chim. Phys. *61*, 826 (1964).
134. *Signorelli, A. J., Hayes, R. G.:* Phys. Rev. *B 8*, 81 (1973).
135. *Bryant, B. W.:* J. Opt. Soc. Am. *55*, 771 (1965).
136. *Jørgensen, C. K.:* In: Gmelins Handbuch der anorganischen Chemie "Seltene Erden", Teil B, Lieferung 1.
137. *Bonnelle, C., Karnatak, R., Jørgensen, C. K.:* Chem. Phys. Letters *14*, 145 (1972).
138. *Jørgensen, C. K.:* Struct. Bonding *22*, 47 (1975).
139. *Callahan, W. R.:* J. Opt. Soc. Am. *53*, 695 (1963).
140. *Jørgensen, C. K., Berthou, H.:* Chem. Phys. Letters *13*, 186 (1972).
141. *Wertheim, G. K., Cohen, R. L., Rosencwaig, A., Guggenheim, H. J.:* Electron spectroscopy (ed. *D. A. Shirley*), p. 813. Amsterdam: North-Holland Publishing Co. 1972.
142. *Herbst, J. F., Lowy, D. N., Watson, R. E.:* Phys. Rev. *B 6*, 1913 (1972).
143. *Ryan, J. L., Jørgensen, C. K.:* J. Phys. Chem. *70*, 2845 (1966).
144. *Suzuki, S., Ishii, T., Sagawa, T.:* Phys. Fennica 9(*S 1*), 310 (1974).
145. *Nagakura, I., Ishii, T., Sagawa, T.:* J. Phys. Soc. Japan *33*, 754 (1952).
146. *Novakov T.:* Phys. Rev. *B 3* 2693 (1971).
147. *Frost, D. C., Ishitani, A., McDowell, C. A.:* Mol. Phys. *24*, 861 (1972).
148. *Rosencwaig, A., Wertheim, G. K., Guggenheim, H. J.:* Phys. Rev. Letters *27*, 479 (1971).

149. *Matienzo L. J., Yin, L. I., Grim, S. O., Swartz, W. E.:* Inorg. Chem. *12*, 2762 (1973).
150. *Yin, L., Adler, I., Tsang, T., Matienzo, L. J., Grim, S. O.:* Chem. Phys. Letters *24*, 81 (1974).
151. *Kim, K. S.:* J. Electron Spectry. *3*, 217 (1974).
152. *Kim, K. S.:* Chem. Phys. Letters *26*, 234 (1974).
153. *Carlson, T. A., Carver, J. C., Saethre, L. J., Garcia Santibánez, F., Vernon, G. A.:* J. Electron Spectry. (Namur 1974 Supplement) *5*, 247 (1974).
154. *Rosencwaig, A., Wertheim, G. K.:* J. Electron Spectry. *1*, 493 (1973).
155. *Hamnett, A.:* Discussion Faraday Soc. *54*, 251 (1973).
156. *Barber, M., Connor, J. A., Hillier, I. H.:* Chem. Phys. Letters *9*, 570 (1971).
157. *Pignataro, S.:* Z. Naturforsch. *27a*, 816 (1972).
158 *Pignataro, S , Foffani, A., Distefano, G.:* Chem. Phys. Letters *20*, 350 (1973).
159. *Wallbank, B., Johnson, C. E., Main, I. G.:* J. Phys. *C*, L 340 and L 493 (1973).
160. *Wertheim, G. K., Rosencwaig, A.:* Phys. Rev. Letters *26*, 1179 (1971).
161. *Spears, D. P., Fischbeck, H. J., Carlson, T. A.:* Phys. Rev. *A 9*, 1603 (1974).
162. *Pollak, R. A., Ley, L., McFeely, F. R., Kowalczyk, S. P., Shirley, D. A.:* J. Electron Spectry. *3*, 381 (1974).
163. *Kowalczyk, S. P., Ley, L., McFeely, F. R., Pollak, R. A., Shirley, D. A.:* Phys. Rev. *B 8*, 3583 (1973).
164. *Barrie, A.:* Chem. Phys. Letters *19*, 109 (1973).
165. *André, J. M., Delhalle, J., Delhalle, S., Caudano, R., Pireaux, J. J., Verbist, J. J.:* Chem. Phys. Letters *23*, 206 (1973).
166. *Escard, J., Mavel, G., Guerchais, J. E., Kergoat, R.:* Inorg. Chem. *14*, 695 (1974).
167. *Jørgensen, C. K.:* Acta Chem. Scand. *16*, 2406 (1962).
168. *Frost, D. C., McDowell, C. A., Woolsey, I. S.:* Chem. Phys. Letters *17*, 320 (1972).
169. *Briggs, D., Gibson, V. A.:* Chem. Phys. Letters *25*, 493 (1974).
170. *Fenger, J., Jørgensen, C. K.:* in preparation.
171. *Wood, P. R., Urch, D. S.:* Phys. Fennica 9(*S 1*), 319 (1974).
172. *Dieke, G. H., Crosswhite, H. M., Dunn, B.:* J. Opt. Soc. Am. *51*, 820 (1961).
173. *Dieke, G. H., Crosswhite, H. M.:* Appl. Opt. *2*, 675 (1963).
174. *Kowalczyk, S. P., Ley, L., Pollak, R. A., McFeely, F. R., Shirley, D. A.:* Phys. Rev. *B 7*, 4009 (1973).
175. *Okusawa, M., Ishii, T., Sagawa, T.:* Physica Fennica 9(*S 1*), 298 (1974).
176. *Bagus, P. S., Freeman, A. J., Sasaki, F.:* Phys. Rev. Letters *30*, 850 (1973).
177. *Combet Farnoux, F.:* Phys. Fennica *9*(*S 1*), 80 (1974).
178. *Bagus, P. S., Wahlgren, U. I.:* Phys. Fennica *9*(*S 1*), 295 (1974).
179. *Viinikka, E. K., Öhrn, Y.:* Phys. Fennica *9*(*S 1*), 304 (1974).
180. *Allen, G. C., Tucker, P. M.:* J. C. S. Dalton *1973*, 470.
181. *Allen, G. C., Crofts, J. A., Curtis, M. T., Tucker, P. M., Chadwick, D., Hampson, P. J.:* J. C. S. Dalton *1974*, 1296.
182. *Keller, C., Engerer, H., Leitner, L., Sriyotha, U.:* J. Inorg. Nucl. Chem. *31*, 965 (1969).
183. *Keller, C., Boroujerdi, A.:* J. Inorg. Nucl. Chem. *34*, 1187 (1972).
184. *Mehlhorn, W.:* Phys. Fennica *9*(*S 1*), 223 (1974).
185. *Wagner, C. D.:* Anal. Chem. *44*, 967 (1972).
186. *Keller, C., Jørgensen, C. K.:* Chem. Phys. Letters *32*, 397 (1975).
187. *Aksela, S., Väyrynen, J., Aksela, H.:* Phys. Rev. Letters *33*, 999 (1974).
188. *Coghlan, W. A., Clausing, R. E.:* A catalog of calculated auger transitions for the elements. ORNL-TM-3576, Oak Ridge 1971.
189. *Packer, M. E., Wilson, J. M.:* Auger transitions. Woking, Surrey: Unwin Brother Ltd. 1973.
190. *Körber, H., Mehlhorn, W.:* Z. Physik *191*, 217 (1966).
191. *Albridge, R. G., Hamrin, K., Johansson, G., Fahlman, A.:* Z. Physik *209*, 419 (1968).
192. *Jørgensen, C. K., Berthou, H.:* Chem. Phys. Letters *25*, 21 (1974).
193. *Mehlhorn, W., Stahlherm, D.:* Z. Physik *217*, 294 (1968).
194. *Aksela, S., Pessa, M., Karras, M.:* Z. Physik *237*, 381 (1970).
195. *Coad, J. P., Rivière, J. C.:* Z. Physik *244*, 19 (1971).
196. *Werme, L. O., Bergmark, T., Siegbahn, K.:* Phys. Scripta (Stockholm) *6*, 141 (1972).

197. *Schröder, J. F., Van Kleef, T. A.:* Physica *49*, 388 (1970).
198. *Powell, C. J., Mandl, A.:* Phys. Rev. Letters *29*, 1153 (1972).
199. *Yin, L., Tsang, T., Adler, I., Yellin, E.:* J. Appl. Phys. *43*, 3464 (1972).
200. *Kowalczyk, S. P., Pollak, R. A., McFeely, F. R., Ley, L., Shirley, D. A.:* Phys. Rev. *B 8*, 2387 (1973).
201. *Schön, G.:* Surface Sci. *35*, 96 (1973).
202. *Allen, G. C., Warren, K. D.:* Inorg. Chem. *8*, 1895 (1969).
203. *Allen, G. C., Warren, K. D.:* Struct. Bonding *9*, 49 (1971).
204. *Schön, G.:* Acta Chem. Scand. *27*, 2623 (1973).
205. *Allen, G. C., Wild, R. K.:* J. C. S. Dalton, 493 (1974).
206. *Al-Joboury, M. I., May, D. P., Turner, D. W.:* J. Chem. Soc. *1965*, 6350.
207. *Ryan, J. L., Jørgensen, C. K.:* Mol. Phys. *7*, 17 (1963).
208. *Bassett, P. J., Lloyd, D. R.:* Chem. Phys Letters *3*, 22 (1969).
209. *Müller, A., Diemann, E., Jørgensen, C. K.:* Struct. Bonding *14*, 23 (1973).
210. *Bird, B. D., Day, P., Grant, E. A.:* J. Chem. Soc. (A) *1970*, 100.
211. *Lempka, H. J., Passmore, T. R., Price, W. C.:* Proc. Roy. Soc. (London) *A 304*, 53 (1968).
212. *Berkowitz, J., Dehmer, J. L., Walker, T. E. H.:* J. Chem. Phys. *59*, 3645 (1973).
213. *Allen, J. D., Boggers, G. W., Goodman, T. D., Wachtel, A. S., Schweitzer, G. K.:* J. Electron Spectry. *2*, 289 (1973).
214. *Goodman, T. D., Allen, J. D., Cusachs, L. C., Schweitzer, G. K.:* J. Electron Spectry. *3*, 289 (1974).
215. *Dehmer, J. L., Berkowitz, J., Cusachs, L. C.:* J. Chem. Phys. *58*, 5681 (1973).
216. *Jørgensen, C. K.:* Mol. Phys. *7*, 417 (1964).
217. *Jørgensen, C. K.:* Acta Chem. Scand. *11*, 166 (1957).
218. *Jørgensen, C. K.:* Halogen chemistry *1*, 265, London: Academic Press 1967.
219. *Jørgensen, C. K.:* Mol. Phys. *2*, 309 (1959).
220. *Cox, P. A., Evans, S., Hamnett, A., Orchard, A. F.:* Chem. Phys. Letters *7*, 414 (1970).
221. *Evans, S., Hamnett, A., Orchard, A. F., Lloyd, D. R.:* Discussions Faraday Soc. *54*, 227 (1973).
222. *Evans, S., Green, J. C., Orchard, A. F., Saito, T., Turner, D. W.:* Chem. Phys. Letters *4*, 361 (1969).
223. *Evans, S., Green, J. C., Jackson, S. E., Hogginson, B.:* J. C. S. Dalton *1974*, 304.
224. *Higginson, B. R., Lloyd, D. R., Burroughs, P., Gibson, D. M., Orchard, A. F.:* J. C. S. Trans. Faraday Soc. II *69*, 1659 (1973).
225. *Nixon, J. F.:* J. C. S. Dalton *1973, 2226.*
226. *Evans, S., Green, J. C., Green, M. L. H., Orchard, A. F., Turner, D. W.:* Discussions Faraday Soc. *47*, 112 (1969).
227. *Evans, S., Green, M. L. H., Jewitt, B., King, G. H., Orchard, A. F.:* J. C. S. Trans. Faraday Soc. II *70*, 356 (1974).
228. *Hillier, I., Guest, M. F., Higgins, B. R., Lloyd, D. R.:* Mol. Phys. *27*, 215 (1974).
229. *Green, J. C., King, D. I., Eland, J. H. D.:* Chem. Commun. *1970*, 1121.
230. *Cocksey, B. G., Eland, J. H. D., Danby, C. J.:* J. C. S. Trans. Faraday Soc. II *69*, 1558 (1973).
231. *Eland, J. H. D.:* Intern. J. Mass. Spectry. Ion Phys. *4*, 37 (1970).
232. *Brundle, C. R., Robin, M. B., Kuebler, N. A., Basch, H.:* J. Am. Chem. Soc. *94*, 1451 (1972).
233. *Jørgensen, C. K.:* Acta Chem. Scand. *12* 903 (1958).
234. *Kono, S., Ishii, T., Sagawa, T., Kobayasi, T.:* Phys. Rev. *B 8*, 795 (1973).
235. *Haas, H., Sheline, R. K.:* J. Am. Chem. Soc. *88*, 3219 (1966).
236. *Wertheim, G. K.. Guggenheim, H. J., Hüfner, S.:* Phys. Rev. Letters *30*, 1050 (1973).
237. *Fadley, C. S., Shirley, D. A.:* J. Res. Natl. Bur. Std. *74A*. 543 (1970).
238. *Baer, Y., Hedén, P. F., Hedman, J., Klasson, M., Nordling, C., Siegbahn, K.:* Phys. Scripta (Stockholm) *1*, 55 (1970).
239. *Hedman, J., Klasson, M., Nilsson, R., Nordling, C., Sorokina, M. F., Kljushnikov, O. I., Nemnonov, S. A., Trapeznikov, V. A., Zyryanov, V. G.:* Phys. Scripta (Stockholm) *4*, 1 (1971).

240. *McFeely, F. R., Kowalczyk, S., Ley, L., Pollak, R. A., Shirley, D. A.:* Phys. Rev. *B 7,* 5228 (1973).
241. *Cavell, R. G., Kowalczyk, S. P., Ley, L., Pollak, R. A., Mills, B., Shirley, D. A., Perry, W.:* Phys. Rev. *B 7,* 5313 (1973).
242. *Vesely, C. J., Kingston, D. L.:* Phys. Rev. *B.8,* 2685 (1973).
243. *Chelikowsky, J., Chadi, D. J., Cohen, M. L.:* Phys. Rev. *B 8,* 2786 (1973).
244. *Shevchik, N. J., Cardona, M., Tejeda, J.:* Phys. Rev. *B 8,* 2833 (1973).
245. *Ley, L., Pollak, R. A., McFeely, F. R., Kowalczyk, S. P., Shirley, D. A.:* Phys. Rev. *B 9,* 600 (1974).
246. *Eastman, D. E., Kuznietz, M.:* J. Appl. Phys. *42,* 1396 (1971).
247. *Li, E. K., Johnson, K. H., Eastman, D. E., Freeouf, J. L.:* Phys. Rev. Letters *32,* 470 (1974).
248. *Wertheim, G. K., Mattheiss, L. F., Campagna, M., Pearsall, T. P.:* Phys. Rev. Letters 32, 997 (1974).
249. *Cox, L. E., Hercules, D. M.:* J. Electron Spectry. *1,* 193 (1973).
250. *Bassett, P. J., Lloyd, D. R.:* J. C. S. Dalton *1972,* 248.
251. *Elbel, S., Bergmann, H., Ensslin, W.:* J. C. S. Faraday Soc. II *70,* 555 (1974).
252. *Jørgensen, C. K.:* Chimia (Aarau) *25,* 109 (1971).
253. *Jørgensen, C. K.:* Chem. Phys. Letters *27,* 305 (1974).
254. *Schäffer, C. E.:* Struct. Bonding *14,* 69 (1973).
255. *Demuynck, J., Veillard, A.:* Theoret. Chim. Acta *28,* 241 (1973).
256. *Bastasz, R., Cahen, D., Lester, J. E., Rajaram, J.:* Chem. Phys. Letters *22,* 489 (1973).
257. *Van der Lugt, W. T. A. M.:* Intern. J. Quantum Chem. *6,* 859 (1972).
258. *Tossell, J. A., Lipscomb, W. N.:* J. Am. Chem. Soc. *94,* 1505 (1972).
259. *Demuynck, J., Veillard, A., Wahlgren, U.:* J. Am. Chem. Soc. *95,* 5563 (1973).
260. *Hand, R. W., Hunt, W. J., Schaefer, H. F.:* J. Am. Chem. Soc. *95,* 4517 (1973).
261. *Dunn, T. M.:* J. Chem. Soc. *1959,* 623.
262. *Williams, R. J. P.:* Discussions Faraday Soc. *26,* 123 (1958).
263. *Wertheim, G. K., Rosencwaig, A., Cohen, R. L., Guggenheim, H. J.:* Phys. Rev. Letters *27,* 505 (1971).
264. *Nugent, L. J., Baybarz, R. D., Burnett, J. L., Ryan, J. L.:* J. Phys. Chem. *77,* 1528 (1973).
265. *Jørgensen, C. K.:* Chimia (Aarau) *23,* 292 (1969).
266. *Jørgensen, C. K.:* Accounts Chem. Res. *4,* 307 (1971).
267. *Linderberg, J., Öhrn, Y.:* Propagators in quantum chemistry. London: Academic Press 1973.
268. *Gelius, U.:* J. Electron Spectry. (Namur 1974 Supplement) *5,* 985 (1974).
269. *Campagna, M., Bucher, E., Wertheim, G. K., Buchanan, D. N. E., Longinotti, L. D.:* Phys. Rev. Letters *32,* 885 (1974).
270. *Carroll, T. X., Shaw, R. W., Thomas, T. D., Kindle, C., Bartlett, N.:* J. Am. Chem. Soc. *96,* 1989 (1974).
271. *Jolly, W. L., Perry, W. B.:* Inorg. Chem. *13,* 2686 (1974).
272. *Ikemoto, I., Ishit, K., Kuroda, H., Thomas, J. M.:* Chem. Phys. Letters *28,* 55 (1974).
273. *Robert, T., Offergeld, G.:* Chem. Phys. Letters *29,* 606 (1974).
274. *Jørgensen, C. K.:* Chimia (Aarau) *29,* 53 (1975).
275. *Kim, K. S., Winograd, N.:* Chem. Phys. Letters *31,* 312 (1975).
276. *Larsson, S.:* Chem. Phys. Letters *32,* 401 (1975).
277. *Asada, S., Satoko, C., Sugano, S.:* J. Phys. Soc. Japan *38,* 855 (1975).
278. *Gupta, R. P., Sen, S. K.:* Bull. Am. Phys. Soc. *20,* 680 (1975).
279. *Wagner, C. D.:* Discussions Faraday Soc. *60,* in press.

Received September 25, 1974

# Fractional Parentage Methods for Ionisation of Open Shells of *d* and *f* Electrons

**P. Anthony Cox**

Équipe de Recherche no. 139 du C.N.R.S., Institut de Chimie, Université Louis Pasteur, BP 296R8, 67–Strasbourg, France

**Table of Contents**

## 1. Introduction

Recent advances in the techniques of photoelectron spectroscopy (*1*) are making it possible to observe ionization from incompletely filled shells of valence elctrons, such as the $3d$ shell in compounds of first-transition-series elements (*2—4*) and the $4f$ shell in lanthanides (*5, 6*). It is certain that the study of such ionisations will give much information of interest to chemists. Unfortunately, however, the interpretation of spectra from open-shell molecules is more difficult than for closed-shell species, since, even in the simple 'one-electron' approach to photoelectron spectra, each orbital shell may give rise to several states on ionisation (*7*). This phenomenon has been particularly studied in the ionisation of core electrons, where for example a molecule (or complex ion in the solid state) with initial spin $S_1$ can generate two distinct states, with spin $S_2 = S_1 - \frac{1}{2}$ or $S_1 + \frac{1}{2}$, on ionisation from a non-degenerate core level (*8*). The analogous effect in valence-shell ionisation was seen by *Wertheim et al.* in the $4f$ band of lanthanide tri-fluorides, $LnF_3$ (*9*). More recent spectra of lanthanide elements and compounds (*6, 9*), show a partial resolution of different orbital states, in addition to spin-multiplicity effects. Different orbital states have also been resolved in gas-phase photoelectron spectra of transition-metal 'sandwich' compounds, such as bis-($\pi$-cyclopentadienyl) complexes (*3, 4*).

A necessary first step in the understanding of spectra such as these, is the purely spectroscopic one of identifying the states which are visible in the spectrum. Only then can chemically interesting information be obtained. It has been suggested (*10, 11*) that a method based on fractional parentage may be able to aid the identification of states, by predicting which states should be visible in the spectrum, and their relative intensities. This method has already been used in certain cases to help interpret spectra (*3, 4, 6, 9*). It must be emphasized that such a theory is based on the same 'one-electron' approximation which predicts a one-one correspondance between photoelectron bands and filled molecular orbitals for closed-shell molecules. Yet in spite of its simple starting point, the fractional-parentage method involves formulae and concepts which are rather unfamiliar to chemical spectroscopists. It therefore seems worthwhile to aid the experimentalists by providing tables of numerical predictions from this theory, covering a range of situations which are likely to be encountered in practice. The publication of such tables, which have already appeared in a thesis (*12*), is the principal aim of this article. At the same time, I have attempted to give a simplified but very general account of the basis of the fractional-parentage theory, to give the formulae which result from this theory (and which have been used in compiling the tables), and to discuss some of the ways in which failures in some of the approximations made in the theory may cause its predictions to break down.

## 2. General Theory

We shall start with the familiar formula for the first-order transition probability between an initial state $|i\rangle$ and a final state $|f\rangle$ (*13*):

$$P_{i \to f} = 2\,\pi/\hbar\;\varrho_f\;|\langle i|\hat{m}|f\rangle|^2 \qquad (1)$$

$\varrho_f$ is the density of final states, which in our case is made up from the degeneracies both of our final ionised state and of the photoelectron. $\hat{m}$ is the electric dipole operator. $|i\rangle$ is the ground state of the molecule, and since we shall be interested in cases where it has both spin and orbital degeneracy, we shall write:

$$|i\rangle = |S_1 A_1 m_1 a_1\rangle \,. \tag{2}$$

The labels show the total spin ($S_1$), the orbital representation ($A_1$), the $z$-component of spin ($m_1$) and the degenerate orbital component ($a_1$). The final state $|f\rangle$ in Eq. (1) represents the ionised state of the molecule, together with the resulting photoelectron, and we shall assume that the matrix element in this equation can be calculated by taking an antisymmetrized product of these two parts:

$$|f\rangle = \mathscr{A}\{|r\ \alpha\ S_2\ A_2\ m_2\ a_2\rangle\ |\chi\rangle\} \,. \tag{3}$$

We have written $|\chi\rangle$ for the continuum function of the photoelectron. The spin and orbital parts of the ionised state function are written ($SAma$) as for the ground state, but we now need two extra labels: $r$ labels the molecular orbital shell which has been ionised, and $\alpha$ takes the value 1, 2, 3... to distinguish between repeated states with the same values of $S_2$ and $A_2$ which may occur in the ionised configuration.

Our intention is not to calculate absolute values for photo-ionisation cross-sections, but rather to relate the cross-sections for the different states which may result from ionising the same orbital shell $r$. Thus we require a formula of the kind:

$$X(r\ \alpha\ S_2\ A_2) = Z(r\ \alpha\ S_2\ A_2 | S_1\ A_1)\ \bar{X}(r) \tag{4}$$

which gives the cross-section $X(r\ \alpha\ S_2\ A_2)$ in terms of some 'average' cross-section for the shell $\bar{X}(r)$ and a numerical coefficient $Z(r\ \alpha\ S_2\ A_2 | S_1\ A_1)$. To make such a factorization, we need to express the matrix element in Eq. (1), which involves states $|i\rangle$ and $|f\rangle$ each having $N$ electrons, in terms of simpler matrix elements involving only the particular orbital which is being ionised. This is easy for a closed-shell molecule, since both the ground and ionised states may be written as single Slater determinants, the ionised state having an electron missing from the shell $r$. Thus if we assume the orbitals to be unchanged on ionisation, Slater's rules for evaluating matrix elements (*14*) give us:

$$\langle i|\hat{m}|f\rangle = \langle \sigma\varrho|\hat{m}|\chi\rangle \tag{5}$$

where ($\sigma\varrho$) labels the particular spin and orbital component of the shell $r$ which has been ionised. To calculate the total cross-section for the ionised state we must sum over its degenerate spin and orbital components. Writing $[r]$ for the orbital degeneracy of the shell $r$ we therefore have

$$X(r\ \alpha\ S_2\ A_2) = 2\,[r]\bar{X}(r) \tag{6}$$

where

$$\bar{X}(r) = \frac{1}{2} \text{ solidus } [r] \sum_{\sigma\varrho} |\langle\sigma\,\varrho\,|\hat{m}|\,\chi\rangle|^2 \tag{7}$$

is the average ionisation cross-section for the shell.

The step expressed by Eq. (5), which is so easy for a closed-shell molecule, is much harder for open-shell systems, because in general neither the ground nor ionised states can be written as single Slater determinants. Instead, we must find suitable linear combinations of such determinants to give the correct spin and symmetry properties of the state (*15, 16*). In simple cases, such linear combinations may be written out explicitly. Thus the matrix element in Eq. (1) could be expanded in terms of matrix elements each involving single Slater determinants. In practice, this method would be very tedious for configurations of $d$ electrons in ligand fields of high symmetry, and impossible for $f$ electrons. The method of fractional parentage may be regarded as a mathematical trick which allows us to short-circuit this lengthy procedure by directly expressing the wave-function for the molecular ground state in terms of the ionised states plus an extra electron.

If we start from the ionised state $|r\,\alpha\,S_2\,A_2\,m_2\,a_2\rangle$ which has $N-1$ electrons, and wish to reconstruct the ground-state $N$-electron wave-function $|S_1\,A_1\,m_1\,a_1\rangle$ by putting back the $N$th electron, we encounter two problems:

(i) We must recover the correct spin and symmetry properties of the ground state;

(ii) The state must be written so as to be antisymmetric with respect to interchange of any two electrons.

The first difficulty means that we must take, not a simple product of one spin-orbital component of the ionised state with one spin-orbital component ($\sigma\varrho$) for the additional electron, but rather a suitable linear combination of such products:

$$\begin{aligned} &|(r\,\alpha\,S_2\,A_2)\,S_1\,A_1\,m_1\,a_1\rangle \\ &\quad = \sum_{\substack{m_2\,a_2\\ \sigma\,\varrho}} \langle S_2\,\tfrac{1}{2}\,m_2\sigma\,|S_1\,m_1\rangle\,\langle A_2\,r\,a_2\varrho\,|A_1\,a_1\rangle \\ &\qquad\qquad |r\,\alpha\,S_2\,A_2\,m_2\,a_2\rangle\,|\sigma\,\varrho\rangle\,. \end{aligned} \tag{8}$$

The left hand side has been written to indicate the parentage of the function: *i.e.* the fact that it has been derived from the ionised state $|r\,\alpha\,S_2\,A_2\rangle$. The first two quantities on the right hand side are coupling coefficients — the linear coefficients required to generate a state with the correct spin and orbital properties (*16, 17*). Although the function written in Eq. (8) has the same spin and symmetry as the ground state, it is clearly not antisymmetric in all $N$ electrons, since the $N$th electron, which is the one added, occupies uniquely the shell $r$. In order to form a properly antisymmetric state, we must ensure that the $N$th electron can occupy all the shells, and also every spin-orbital component of each shell. This can only be done by taking a suitable linear combination of functions like that in (8), formed from all possible ion-state 'parents':

$$|i\rangle = \sum_{r\,\alpha\,S_2\,A_2} Q(r\,\alpha\,S_2\,A_2)\;|(r\,\alpha\,S_2\,A_2)\,S_1\,A_1\,m_1\,a_1\rangle\,. \tag{9}$$

The coefficients $Q(r\,\alpha\,S_2\,A_2)$ are chosen to make the state $|i\rangle$ properly antisymmetric, and are the *coefficients of fractional parentage.*

The final step before we can simplify the $N$-electron matrix in Eq. (1) is to remove the antisymmetrization of the final state by writing:

$$\langle i|\hat{m}|f\rangle = N^{\frac{1}{2}}\,\langle i|\hat{m}|\,\{|r\,\alpha\,S_2\,A_2\,m_2\,a_2\rangle\,|\chi(N)\rangle\}\,. \tag{10}$$

On the right hand side, the ion state is still antisymmetric in the first $N-1$ electrons, but the continuum orbital is occupied exclusively by the $N$th. That it is possible to write this may be seen by expanding the left hand side as a sum of $N$ terms, in which the continuum orbital is occupied by a different electron in turn. Each term must have the same value, since all electrons are equivalent in the state $|i\rangle$. Thus we have $N$ equal terms, with a normalization factor of $N^{-\frac{1}{2}}$.

Replacing the ground state $|i\rangle$ in Eq. (10) by its expansion according to Eqs. (8) and (9), and using the fact that all the ionised states are orthogonal, we find:

$$\begin{aligned}\langle i|\hat{m}|f\rangle = {}& N^{\frac{1}{2}}\,Q(r\,\alpha\,S_2\,A_2)\\ & \sum_{\sigma\,\varrho} \langle S_2\,\tfrac{1}{2}\,m_2\sigma|S_1\,m_1\rangle\,\langle A_2\,r\,a_2\varrho|A_1\,a_1\rangle\,\langle\sigma\,\varrho|\hat{m}|\chi\rangle\end{aligned} \tag{11}$$

thus effecting the desired reduction to a one-electron matrix element. We must now square (11), sum over the degenerate components $(m_2\,a_2)$ of the ionised state, and average over the ground state components $(m_1\,a_1)$. The coupling coefficients have the following useful orthonormality properties:

$$\sum_{m_1\,m_2} \langle S_2\,\tfrac{1}{2}\,m_2\,\sigma\,|\,S_1\,m_1\rangle\,\langle S_2\,\tfrac{1}{2}\,m_2\,\sigma'\,|S_1\,m_1\rangle = \delta_{\sigma\sigma'}\,(2S_1+1)/2 \tag{12a}$$

$$\sum_{a_1\,a_2} \langle A_2\,r\,a_2\varrho\,|\,A_1\,a_1\rangle\,\langle A_2\,r\,a_2\,\varrho'\,|A_1\,a_1\rangle = \delta_{\varrho\varrho'}\,[A_1]/[r] \tag{12b}$$

where $[r]$ and $[A_1]$ represent, respectively, the orbital degeneracies of the shell $r$, and the ground state *(16, 17)*. Thus:

$$\begin{aligned}&\frac{1}{(2S_1+1)\,[A_1]}\sum_{m_1 m_2 a_1 a_2} |\langle i|\hat{m}|f\rangle|^2\\ &= N\,Q^2(r\,\alpha\,S_2\,A_2)\,\frac{1}{2[r]}\sum_{\sigma\,\varrho} |\langle\sigma\,\varrho|m|\chi\rangle|^2\,.\end{aligned} \tag{13}$$

The last term in (13) is the average one-electron cross-section defined in Eq. (7) and thus we obtain a formula for the coefficients required in Eq. (4):

$$Z(r \alpha S_2 A_2 | S_1 A_1) = N Q^2(r \alpha S_2 A_2) . \quad (14)$$

This general formula allows us to treat a wide range of special cases, for which formulae are given in the next section. The derivation of such formulae is a purely mathematical problem, which need not concern us here. The essential physics is contained in the above discussion, which also brings out an important point: no assumptions have been made as to the polarization of the electromagnetic radiation, or the direction of the photoelectron. Thus the formulae should be valid for substances in the solid, as in the gas phase, and should be quite independent of the angle at which photoelectrons are observed.

## 3. Basic Intensity Rules

The intensity formula, Eq. (14), which was derived in the last section, is too general for practical purposes, and so in this section we shall quote some useful consequences of it, and some formulae applicable to special cases.

(i) Selection Rules:

The intensity of the ionised state is zero unless:

$$|S_1 - S_2| = \tfrac{1}{2} \quad (15)$$

and:

$$A_1 \otimes A_2 \supset r . \quad (16)$$

That is: the total spin must change by $\pm\frac{1}{2}$, and the direct product of the orbital representations of the final and initial states must contain the orbital representation of the ionised shell. For atoms, $A_1$, $A_2$ and $r$ would be replaced by orbital angular momenta $L_1$, $L_2$ and $l$ respectively, and Eq. (16) written:

$$|L_1 - L_2| \leqslant l . \quad (17)$$

(ii) Ionisation of a single open shell (*10*):

If the ionised shell is the only open shell in the molecule:

$$Z(r \alpha S_2 A_2 | S_1 A_1) = n_r \, |\langle \alpha S_2 A_2 \cdot r | \} S_1 A_1 \rangle|^2 \quad (18)$$

where $n_r$ is the number of electrons initially in the shell, and is multiplied by the square of the single-shell fractional parentage coefficient (*18, 19*). Values of these coefficients have been calculated and tabulated for configurations of electrons in atomic (*20*) and molecular (*16*) orbitals.

(iii) Ionisation of a closed shell *(10)*:

If a closed shell is ionised, all possible states resulting from spin and orbital coupling of the hole with the open shell will be produced. The total intensity of ionised states with spin and symmetry $(S_2 A_2)$ is proportional to the spin-orbital degeneracy:

$$\sum_{\alpha} Z(r\,\alpha\, S_2\, A_2 | S_1\, A_1) = \frac{(2S_2 + 1)\,[A_2]}{(2S_1 + 1)\,[A_1]} \,. \tag{19}$$

The form of this equation is intended to emphasize that more than one state $(S_2 A_2)$ may appear, in which case the total intensity given by (19) is shared between them. This phenomenon, which has been observed for example in the photoelectron spectrum of $O_2$ *(21)*, is discussed in Section 5.

(iv) Several open shells *(11)*:

When the ionised shell $r$ is one of several open shells in the molecule (as may happen with configurations of $e$ and $t_2$ electrons in cubic ligand fields), more complicated formulae must be applied. Suppose that the ground state has two open shells $r$ and $s$ in which the electrons are separately coupled to form functions with spin and symmetry $(S_3 A_3)$ and $(S_4 A_4)$ respectively. Thus we write this state:

$$|(S_3\, A_3)\, (S_4\, A_4)\, S_1\, A_1\, m_1\, a_1\rangle \tag{20}$$

to indicate the coupling scheme. On ionisation from the shell $r$, the function $(S_3 A_3)$ changes to $(S_5 A_5)$, whereas the function for the other shell remains unchanged. Then:

$$\begin{aligned} Z(r\,\alpha\, S_2\, A_2 | S_1\, A_1) &= n_r (2S_2 + 1)\,(2S_3 + 1)\,[A_2]\,[A_3] \\ &\times |\langle S_5\, A_5 \cdot r|\} S_3\, A_3\rangle|^2\, W^2(S_5\, S_3\, S_2\, S_1\,;\, \tfrac{1}{2} S_4) \\ &\times W^2(A_5\, A_3\, A_2\, A_1\,;\, r A_4)\,. \end{aligned} \tag{21}$$

As before $[A]$ indicates orbital degeneracy; Eq. (21) also involves $W$ or Racah coefficients for spin and orbital coupling. Values for the spin coefficients may be calculated from formulae quoted by *Brink* and *Satchler* *(17)*. Frequently, however, it may not be necessary to use this formula, since it may be clear from the selection rules that a given function $(S_5 A_5)$ of the ionised shell can only produce one allowed resultant state $(S_2 A_2)$, and in this case the intensity is entirely determined by the fractional parentage coefficient for the ionised shell:

$$Z(r\,\alpha\, S_2\, A_2 | S_1\, A_1) = n_r |\langle S_5\, A_5 \cdot r|\}\, S_3\, A_3\rangle|^2 \,. \tag{22}$$

One problem with the application of these formulae is that states of the same symmetry, either for the neutral of ionised molecule, may arise from different combinations of separate functions for the shells $r$ and $s$. Thus we may have intra-configurational mixing effects which mean that the simple break-down of the states assumed here is not valid. This point is discussed at greater length in Section 5.

(v) Sum rule for shells (*10*):

Summing the intensity over all states which can be formed by ionising the shell $r$, we have:

$$\sum_{\alpha S_2 A_2} Z(r\,\alpha\,S_2\,A_2|S_1\,A_1) = n_r\,. \tag{23}$$

Thus the total intensity from a shell, whether it is open or closed, is determined by the number of electrons in the shell.

(vi) Sum rule for states of a given spin multiplicity (*12*):

Instead of summing over all the state of the ionised configuration, we may sum over all orbital states which have a given spin multiplicity, thus obtaining formulae which are useful in situations where the individual orbital states are not resolved. The final spin $S_2$ can only take the two values $S_1 - \frac{1}{2}$ and $S_1 + \frac{1}{2}$, and we have:

$$\sum_{\alpha A_2} Z\left(r\,\alpha\,S_1 - \tfrac{1}{2}\,A_2|S_1\,A_1\right) = \frac{S_1\,(n_r + 2S_1 + 2)}{2S_1 + 1} \tag{24a}$$

$$\sum_{\alpha A_2} Z\left(r\,\alpha\,S_1 + \tfrac{1}{2}\,A_2|S_1\,A_1\right) = \frac{(S_1 + 1)\,(n_r - 2S_1)}{2S_1 + 1}\,. \tag{24b}$$

## 4. Spin-Orbit Coupling

In the (*j—j*) limit of strong spin-orbit coupling (*15*) it would still be possible to use formulae (18) and (19) of the previous section, if each spin-orbit component of a given orbital shell were itself regarded as a separate shell. The labels ($S_1\,A_1$) and ($S_2\,A_2$) would be replaced by double-group labels ($B_1$) and ($B_2$) for the spin-orbit states. Thus, for ionisation from a single open shell. we could write:

$$Z(r\,\alpha\,B_2|B_1) = n_r|\langle\alpha\,B_2\cdot r|B_1\rangle|^2 \tag{25}$$

using a fractional parentage coefficient for (*j—j*) coupling (*22*). Eq. (19), for the ionization of a closed shell, would become:

$$\sum_{\alpha} Z(r\,\alpha\,B_2|B_1) = [B_2]/[B_1]\,. \tag{26}$$

It will be more useful in practice, however, to discuss the Russell-Saunders limit of weak spin-orbit coupling (*15, 16*). We shall show how to derive formulae applicable to atoms in spherical symmetry. Thus we shall change the notation to that used in atomic spectroscopy (*15*), writing $|S\,L\,M_S\,M_L\rangle$ for the ground state and $|l\,\alpha\,S'\,L'\,M_S'\,M_L'\rangle$ for an ionised state. With spin-orbit coupling we must take linear combinations of different $(M_S\, M_L)$ combinations for a particular $J$ value:

$$|S\, L\, J\, M_J\rangle = \sum_{M_S M_L} \langle S\, L\, M_S\, M_L | J\, M_J\rangle\, |S\, L\, M_S\, M_L\rangle \qquad (27a)$$

$$|l\, \alpha\, S'\, L'\, J'\, M'_J\rangle = \sum_{M'_S M'_L} \langle S'\, L'\, M'_S\, M'_L\, | J'\, M'_J\rangle\, |l\, \alpha\, S'\, L'\, M'_S\, M'_L\rangle\,. \qquad (27b)$$

We can combine these equations for the intial and ionised states with Eqs. (8) to (10) of Section 2, to obtain the following expression for the photoionisation matrix element:

$$\begin{aligned}\langle i|\hat{m}|f\rangle = N^{\frac{1}{2}}\, Q(l\, \alpha\, S'\, L') \sum_{M_S M'_S m_s} \sum_{M_L M'_L m'_l} \{\langle S\, L\, M_S\, M_L | J\, M_J\rangle \\ \langle S'\, L'\, M'_S\, M'_L | J'\, M'_J\rangle \langle S'\, \tfrac{1}{2}\, M'_S\, m'_s | S\, M_S\rangle \langle L'\, l\, M'_L\, m'_l\, | L\, M_L\rangle \\ \langle m_s\, m_l | \hat{m} | \chi\rangle\}\,.\end{aligned} \qquad (28)$$

This rather unpleasant expression must be squared, and then summed over the degenerate components $M'_J$ of the final state, and averaged over the ground-state components $M_J$. The resulting product of coupling coefficients can be evaluated by the methods described by *Brink* and *Satchler* (*17*), and the final result may be written:

$$\begin{aligned}Z(l\, \alpha\, S'\, L'\, J' | S\, L\, J) = Z(l\, \alpha\, S'\, L' | S\, L) \\ \times\, (2J' + 1)\, (2S + 1)\, (2L + 1) \sum_j (2j + 1) \begin{pmatrix} \tfrac{1}{2} & l & j \\ S & L & J \\ S' & L' & J' \end{pmatrix}^2\end{aligned} \qquad (29)$$

The first term on the right hand side is simply the relative intensity of the state $|l\, \alpha\, S'\, L'\rangle$ without spin-orbit coupling. The formula then gives the intensity of the ion multiplet component $J'$ starting from a component $J$ of the ground state, and involves a 9—$j$ coefficient, which may be evaluated from standard formulae (*17*). The summation is over the two values of $j = l - \frac{1}{2}$ and $l + \frac{1}{2}$.

Apart from its use for quantitative calculations, Eq. (29) has a number of interesting consequences:

(i) Selection rule:

$$|J' - J| \leqslant l + \tfrac{1}{2}\,. \qquad (30)$$

(ii) Sum rule:

The sum over the different components $J'$ of the ionised state gives the intensity calculated without spin-orbit coupling. Thus spin-orbit coupling can be neglected unless it is resolved in the spectrum, even if the ground state of the molecule is significantly split, so that only one value of $J$ is populated.

(iii) Average over ground-state components:

If the ground state is an orbital singlet, and so only has one multiplet component $J$, or if the ground-state multiplet splitting is so small that multiplet components are thermally occupied in proportion to their degeneracies, the spin-orbit factor in Eq. (29) becomes simply:

$$\frac{(2J' + 1)}{(2S' + 1)\,(2L' + 1)} \,. \tag{31}$$

## 5. Applicability of the Theory

It is important to recognise that the theory presented here is based on a large number of approximations, and it may be worthwhile at this point to examine some of these, so as to see under what conditions the predictions of the theory can be relied upon.

(i) By using a simple first-order formula for the transition probability [Eq. (1)], we are neglecting auto-ionisation effects. These effects can be extreemly important close to the ionisation threshold (*23*), where they may upset the pattern of intensities. They are not likely to be important, however, when the photoelectron kinetic energy is more than a few electron-volts.

(ii) Factorising the continuum function for the photoelectron out of the final state [Eq. (3)] neglects correlation effects between the out-going electron and the ionised molecular state. Such correlation seems to be important in determining the probability of two-electron 'shake-off' processes (*24*), but these is no evidence that it has a large influence (except in the case of auto-ionisation mention above) on the cross-section for one-electron ionisation.

(iii) The most serious approximation made is undoubtably the neglect both of correlation (or configuration-interaction) effects in the neutral and ionised states of the molecule, and of the relaxation of molecular orbitals which occurs on ionisation. It is well known from the spectra of closed-shell molecules that these effects can cause 'forbidden' shake-up states to appear in the spectrum, particularly in core ionisation (*25*). In the valence region, shake-up states are generally found to be weak, and considerably higher in energy than most of the interesting primary ionised states (*26*). This suggests that the approximation used here may often be quite good for determining intensities in the valence region. However, configuration-interaction effects have been shown to be important in the spectrum of nitric oxide (*27*), and this may also prove to be the case with transition-metal compounds. Another cause for concern is that molecular orbital calculations have shown relaxation effects to be especially large for ionisation of orbitals with a large amount in transition-metal 3d character (*28*). More data are required before it becomes clear how far these effects may upset the predictions of the present theory: in the series of compounds which has been most completely studied so far, that of bis-($\pi$-cyclopentadienyl) complexes of transition metals, the theoretical predictions seem to be quite well obeyed, although some uncertainties in interpretation remain (*3*).

(iv) We have assumed that the average cross-section $\bar{X}(r)$ of Eq. (4) will be the same for all states formed by ionising the same molecular orbital shell. This will not be true if the energy separation between the states is an appreciable fraction of the photoelectron kinetic energy, since both the matrix element and the density of final states in Eq. (1) depend on this kinetic energy. It is sometimes assumed (*29*) that $\bar{X}(r)$ will be approximately the same for different shells, so that in a closed-shell molecule the intensity of ionisation will reflect the orbital degeneracy of the shell. This assumption is certainly not correct, however, if the molecular orbitals concerned have appreciably different atomic orbital composition, for example in the metallocenes, studied with He(I) radiation, (*3*) different cross-sections are found not only for ligand and metal orbitals, but also for the different metal orbitals ($a_1$ and $e_2$). In Tables 1 and 2, which give the predicted intensities for *d* electrons in ligand fields, care has been taken to indicate which orbital is being ionised (for example; *e* or $t_2$ for cubic fields) so that some allowance may be made for this effect.

(v) In parts (iii) and (iv) of Section 3, we noted the difficulties which may arise when several states of the same spin and symmetry are possible in the same ionised configuration. It will be best to illustrate these by means of a specific example. Consider the ground state ($t_2^6\,e^2$) ${}^3A_2$ which would arise from a $d^8$ system in an octahedral ligand field. Ionisation of an *e* electron can only produce the state ${}^2E$, but on ionisation of a $t_2$ electron, the resulting 'hole' can couple with the ${}^3A_2$ state of the open *e* shell, to produce two states, ${}^4T_1$ and ${}^2T_1$. Eq. (19) shows that the intensity factors for these states will be 4 and 2 respectively. However, the ionised configuration $t_2^5\,e^2$ gives rise to a second ${}^2T_1$ state, by coupling the $t_2$ hole with the ${}^1E$ state of the $e^2$ configuration. We might say that this other ${}^2T_1$ state would be forbidden in ionisation, since the electrons in the *e* shell would have to rearrange from their initial state ${}^3A_2$. Unfortunately, however, the actual ${}^2T_1$ states which occur in nature will be mixtures of the ones considered here, and so in practice both states will appear in the spectrum. Their relative intensity will depend on the proportion of the allowed state they have in them, which can only be calculated by diagonalising the matrix of electrostatic interaction (*16*). Thus in this, and in related cases, the fractional parentage method can only predict a total intensity which may be distributed between two or more states. Similar difficulties occur in ionisation of *f* electrons: thus ionisation from the ${}^4I$ ground state of $f^{11}$ can produce no less than four ${}^3$H states: again, the relative intensities of these can only be determined by finding the eigenvectors of an electrostatic matrix.

## 6. Tables

The following tables have been compiled using the formulae given in Sections 3 and 4. The numbers given are in fact the *Z* coefficients defined in Eq. (4), and are thus the intensities expressed in units of the one-electron cross-section for the orbital which is ionised. Cases where repeated states can arise, as discussed above, are marked with an asterix (*): in these cases the total intensity given in the table may be distributed among two or more states.

Tables 1 and 2 assume the 'strong-field' limit for configurations of $d$ electrons in ligand fields (*16*). In Table 1, all ground states have been considered which are likely to occur in tetrahedral and octahedral symmetries. Table 2 was originally intended for $D_5$ symmetries appropriate to the metallocenes, but it can also be used for $D_n$ where $n > 5$, and (with some change of notation) for linear symmetries.

Table 3 shows how the intensity of ionisation from an open-shell configuration is divided between states of the two possible spin multiplicities: see Eq. (24).

Table 4 shows the predicted intensities from $f^n$ configurations: in the final column the expected distribution among spin-orbit components is presented. It has been assumed that only the lowest $J$ component of the ground state is appreciably populated. Since the predicted distribution is very uneven, some allowed components having a normalised intensity less than $10^{-4}$, only those components for which the intensity is $\geqslant 0.1$ have been listed. In repeated (LS) states, for which only summed intensities are listed, the relative strengths for different $J$ values should be the same.

Unfortunately Eq. (29), which gives the intensities with spin-orbit coupling, is not directly applicable to molecular symmetry groups. However, for the most interesting case, that of $t_2^n$ configurations in octahedral fields, the well known isomorphism with $p^n$ configurations can be used (*16*). The resulting predictions are shown in Table 5, with those for $(j\text{—}j)$ coupling in Table 6. The notation for double-group representations follows (*16*).

Table 1. States arising in ionisation from $d^n$ configurations in cubic fields

| Initial configuration and state | | | Electron ionised | Final states | Intensities |
|---|---|---|---|---|---|
| $d^2$ | $t_2^2$ | $^3T_1$ | $t_2$ | $^2T_2$ | 2 |
| | $e^2$ | $^3A_2$ | $e$ | $^2E$ | 2 |
| $d^3$ | $t_2^3$ | $^4A_2$ | $t_2$ | $^3T_1$ | 3 |
| | $e^2t_2$ | $^1T_1$ | $t_2$ | $^3A_2$ | 1 |
| | | | $e$ | $^3T_1$ | 1 |
| | | | | $^3T_2$ | 1 |
| | $e^3$ | $^2E$ | $e$ | $^3A_2$ | $^3/_2$ |
| | | | | $^1A_1$ | $^1/_2$ |
| | | | | $^1E$ | 1 |
| $d^4$ | $t_2^4$ | $^3T_1$ | $t_2$ | $^4A_2$ | $^4/_3$ |
| | | | | $^2E$ | $^2/_3$ |
| | | | | $^2T_1$ | 1 |
| | | | | $^2T_2$ | 1 |
| | $t_2^3e$ | $^5E$ | $e$ | $^4A_2$ | 1 |

Table 1 (continued)

| Initial configuration and state | | | Electron ionised | Final states | Intensities |
|---|---|---|---|---|---|
| $d^4$ (cont'd) | | | | | |
| | | | $t_2$ | $^4T_1$ | $^3/_2$ |
| | | | | $^4T_2$ | $^3/_2$ |
| | $e^2t_2^2$ | $^5T_2$ | $t_2$ | $^4T_1$ | 2 |
| | | | $e$ | $^4T_1$ | 1 |
| | | | | $^4T_2$ | 1 |
| | $e^4$ | $^1A_1$ | $e$ | $^2E$ | 4 |
| $d^5$ | $t_2^3e^2$ | $^6A_1$ | $e$ | $^5E$ | 2 |
| | | | $t_2$ | $^5T_2$ | 3 |
| | $t_2^5$ | $^2T_2$ | $t_2$ | $^3T_1$ | 3 |
| | | | | $^1A_1$ | $^1/_3$ |
| | | | | $^1E$ | $^2/_3$ |
| | | | | $^1T_2$ | 1 |
| | $e^4t_2$ | $^2T_2$ | $t_2$ | $^1A_1$ | 1 |
| | | | $e$ | $^3T_1$ | $^3/_2$ |
| | | | | $^1T_1$ | $^1/_2$ |
| | | | | $^3T_2$ | $^3/_2$ |
| | | | | $^1T_2$ | $^1/_2$ |
| $d^6$ | $t_2^6$ | $^1A_1$ | $t_2$ | $^2T_2$ | 6 |
| | $t_2^4e^2$ | $^5T_2$ | $e$ | $^4T_1$ | 1 |
| | | | | $^4T_2$ | 1 |
| | | | $t_2$ | $^6A_1$ | $^6/_5$ |
| | | | | $^4A_1$ | $^2/_{15}$ |
| | | | | $^4E$ | $^2/_3$* |
| | | | | $^4T_1$ | 1 |
| | | | | $^4T_2$ | 1 |
| | $e^3t_2^3$ | $^5E$ | $t_2$ | $^4T_1$ | $^3/_2$ |
| | | | | $^4T_2$ | $^3/_2$ |
| | | | $e$ | $^6A_1$ | $^6/_5$ |
| | | | | $^4A_1$ | $^3/_{10}$ |
| | | | | $^4A_2$ | $^1/_2$ |
| | | | | $^4E$ | 1* |
| $d^7$ | $t_2^6e$ | $^2E$ | $e$ | $^1A_1$ | 1 |
| | | | $t_2$ | $^3T_1$ | $^9/_4$ |
| | | | | $^3T_2$ | $^9/_4$ |
| | | | | $^1T_1$ | $^3/_4$ |
| | | | | $^1T_2$ | $^3/_4$ |

Table 1 (continued)

| Initial configuration and state | | | Electron ionised | Final states | Intensities |
|---|---|---|---|---|---|
| $d^7$ (cont'd) | | | | | |
| | $t_2^5e^2$ | $^4T_1$ | $e$ | $^3T_1$ | 1 |
| | | | | $^3T_2$ | 1 |
| | | | $t_2$ | $^5T_2$ | $^5/_2$ |
| | | | | $^3A_2$ | $^1/_3$ |
| | | | | $^3E$ | $^2/_3$ |
| | | | | $^3T_1$ | 1 |
| | | | | $^3T_2$ | $^1/_2$* |
| | $e^4t_2^3$ | $^4A_2$ | $t_2$ | $^3T_1$ | 3 |
| | | | $e$ | $^5E$ | $^5/_2$ |
| | | | | $^3E$ | $^3/_2$* |
| $d^8$ | $t_2^6e^2$ | $^3A_2$ | $e$ | $^2E$ | 2 |
| | | | $t_2$ | $^4T_1$ | 4 |
| | | | | $^2T_1$ | 2* |
| | $e^4t_2^4$ | $^3T_1$ | $t_2$ | $^4A_2$ | $^4/_3$ |
| | | | | $^2E$ | $^2/_3$ |
| | | | | $^2T_1$ | 1 |
| | | | | $^2T_2$ | 1 |
| | | | $e$ | $^4T_1$ | $^4/_3$ |
| | | | | $^4T_2$ | $^4/_3$ |
| | | | | $^2T_1$ | $^2/_3$* |
| | | | | $^2T_2$ | $^2/_3$* |
| $d^9$ | $t_2^6e^3$ | $^2E$ | $e$ | $^3A_2$ | $^3/_2$ |
| | | | | $^1A_1$ | $^1/_2$ |
| | | | | $^1E$ | 1 |
| | | | $t_2$ | $^3T_1$ | $^9/_4$ |
| | | | | $^3T_2$ | $^9/_4$ |
| | | | | $^1T_1$ | $^3/_4$ |
| | | | | $^1T_2$ | $^3/_4$ |
| | $e^4t_2^5$ | $^2T_2$ | $t_2$ | $^3T_1$ | 3 |
| | | | | $^1A_1$ | $^1/_3$ |
| | | | | $^1E$ | $^2/_3$ |
| | | | | $^1T_2$ | 1 |
| | | | $e$ | $^3T_1$ | $^3/_2$ |
| | | | | $^1T_1$ | $^1/_2$ |
| | | | | $^3T_2$ | $^3/_2$ |
| | | | | $^1T_2$ | $^1/_2$ |

Table 2. States arising in ionisation from $d^n$ configurations in fields of $D_5$ symmetry

| Initial configuration and state | | | Electron ionised | Final states | Intensities |
|---|---|---|---|---|---|
| $d^2$ | $a_1^2$ | $^1A_1$ | $a_1$ | $^2A_1$ | 2 |
| | $e_2^2$ | $^3A_2$ | $e_2$ | $^2E_2$ | 2 |
| | $a_1e_2$ | $^3E_2$ | $e_2$ | $^2A_1$ | 1 |
| | | | $a_1$ | $^2E_2$ | 1 |
| $d^3$ | $a_1^2e_2$ | $^2E_2$ | $e_2$ | $^1A_1$ | 1 |
| | | | $a_1$ | $^3E_2$<br>$^1E_2$ | $^3/_2$<br>$^1/_2$ |
| | $a_1e_2^2$ | $^4A_2$ | $e_2$ | $^3E_2$ | 2 |
| | | | $a_1$ | $^3A_2$ | 1 |
| | $e_2^3$ | $^2E_2$ | $e_2$ | $^3A_2$<br>$^1A_1$<br>$^1E_1$ | $^3/_2$<br>$^1/_2$<br>1 |
| $d^4$ | $a_1^2e_2^2$ | $^3A_2$ | $e_2$ | $^2E_2$ | 2 |
| | | | $a_1$ | $^4A_2$<br>$^2A_2$ | $^4/_3$<br>$^2/_3$ |
| | $a_1e_2^3$ | $^3E_2$ | $e_2$ | $^4A_2$<br>$^2A_1$<br>$^2A_2$<br>$^2E_1$ | $^4/_3$<br>$^1/_2$<br>$^1/_6$<br>1 |
| | | | $a_1$ | $^2E_2$ | 1 |
| | $a_1e_2^2e_1$ | $^5E_1$ | $e_1$ | $^4A_2$ | 1 |
| | | | $e_2$ | $^4E_1$<br>$^4E_2$ | 1<br>1 |
| | | | $a_1$ | $^4E_1$ | 1 |
| $d^5$ | $a_1^2e_2^3$ | $^2E_2$ | $e_2$ | $^3A_2$<br>$^1A_1$<br>$^1E_1$ | $^3/_2$<br>$^1/_2$<br>1 |
| | | | $a_1$ | $^3E_2$<br>$^1E_2$ | $^3/_2$<br>$^1/_2$ |
| | $a_1e_2^4$ | $^2A_1$ | $e_2$ | $^3E_2$<br>$^1E_2$ | 3<br>1 |
| | | | $a_1$ | $^1A_1$ | 1 |

Table 2 (continued)

| Initial configuration and state | | | Electron ionised | Final states | Intensities |
|---|---|---|---|---|---|
| $d^5$ (cont'd) | | | | | |
| | $a_1e_2^2e_1^2$ | $^6A_1$ | $e_1$ | $^5E_1$ | 2 |
| | | | $e_2$ | $^5E_2$ | 2 |
| | | | $a_1$ | $^5A_1$ | 1 |
| | $a_1^2e_2^2e_1$ | $^4E_1$ | $e_1$ | $^3A_2$ | 1 |
| | | | $e_2$ | $^3E_1$<br>$^3E_2$ | 1<br>1 |
| | | | $a_1$ | $^5E_1$<br>$^3E_1$ | $^5/_4$<br>$^3/_4$* |
| $d^6$ | $a_1^2e_2^4$ | $^1A_1$ | $e_2$ | $^2E_2$ | 4 |
| | | | $a_1$ | $^2A_1$ | 2 |
| | $a_1^2e_2^2e_1^2$ | $^5A_1$ | $e_1$ | $^4E_1$ | 2 |
| | | | $e_2$ | $^4E_2$ | 2 |
| | | | $a_1$ | $^6A_1$<br>$^4A_1$ | $^6/_5$<br>$^4/_5$ |
| | $a_1e_2^4e_1$ | $^3E_1$ | $e_1$ | $^2A_1$ | 1 |
| | | | $e_2$ | $^4E_1$<br>$^4E_2$<br>$^2E_1$<br>$^2E_2$ | $^4/_3$<br>$^4/_3$<br>$^2/_3$<br>$^2/_3$ |
| | | | $a_1$ | $^2E_1$ | 1 |
| | $a_1e_2^3e_1^2$ | $^5E_2$ | $e_1$ | $^4E_1$<br>$^4E_2$ | 1<br>1 |
| | | | $e_2$ | $^6A_1$<br>$^4A_1$<br>$^4A_2$<br>$^4E_1$ | $^6/_5$<br>$^3/_{10}$*<br>$^1/_2$*<br>1 |
| | | | $a_1$ | $^4E_2$ | 1 |
| $d^7$ | $a_1^2e_2^4e_1$ | $^2E_1$ | $e_1$ | $^1A_1$ | 1 |
| | | | $e_2$ | $^3E_1$<br>$^3E_2$<br>$^1E_1$<br>$^1E_2$ | $^3/_2$<br>$^3/_2$<br>$^1/_2$<br>$^1/_2$ |

Table 2 (continued)

| Initial configuration and state | | | Electron ionised | Final states | Intensities |
|---|---|---|---|---|---|
| $d^7$ (cont'd) | | | | | |
| | | | $a_1$ | $^3E_1$ | $^3/_2$ |
| | | | | $^1E_1$ | $^1/_2$ |
| | $a_1^2e_2^3e_1^2$ | $^4E_2$ | $e_1$ | $^3E_1$ | 1 |
| | | | | $^3E_2$ | 1 |
| | | | $e_2$ | $^5A_1$ | $^5/_4$ |
| | | | | $^3A_1$ | $^1/_4$ |
| | | | | $^3A_2$ | $^1/_2$* |
| | | | | $^3E_1$ | 1 |
| | | | $a_1$ | $^5E_2$ | $^5/_4$ |
| | | | | $^3E_2$ | $^3/_4$* |
| | $a_1e_2^4e_1^2$ | $^4A_2$ | $e_1$ | $^3E_1$ | 2 |
| | | | $e_2$ | $^5E_2$ | $^5/_2$ |
| | | | | $^3E_2$ | $^3/_2$* |
| | | | $a_1$ | $^3A_2$ | 1 |
| $d^8$ | $a_1^2e_2^4e_1^2$ | $^3A_2$ | $e_1$ | $^2E_1$ | 2 |
| | | | $e_2$ | $^4E_2$ | $^8/_3$ |
| | | | | $^2E_2$ | $^4/_3$* |
| | | | $a_1$ | $^4A_1$ | $^4/_3$ |
| | | | | $^2A_1$ | $^2/_3$* |
| $d^9$ | $a_1^2e_2^4e_1^3$ | $^2E_1$ | $e_1$ | $^3A_2$ | $^3/_2$ |
| | | | | $^1A_1$ | $^1/_2$ |
| | | | | $^1E_2$ | 1 |
| | | | $e_2$ | $^3E_1$ | $^3/_2$ |
| | | | | $^3E_2$ | $^3/_2$ |
| | | | | $^1E_1$ | $^1/_2$ |
| | | | | $^1E_2$ | $^1/_2$ |
| | | | $a_1$ | $^3E_1$ | $^3/_2$ |
| | | | | $^1E_1$ | $^1/_2$ |

Table 3. Total intensities of states of a given multiplicity arising in ionisation from a configuration

| Initial configuration | | Total intensities of multiplicities | |
|---|---|---|---|
| Number of electrons | Multiplicity | Higher | Lower |
| 3 | 2 | $^3/_2$ | $^3/_2$ |
| 4 | 3 | $^4/_3$ | $^8/_3$ |
| 5 | 2 | 3 | 2 |
| | 4 | $^5/_4$ | $^{15}/_4$ |
| 6 | 3 | $^8/_3$ | $^{10}/_3$ |
| | 5 | $^6/_5$ | $^{24}/_5$ |
| 7 | 2 | $^9/_2$ | $^5/_2$ |
| | 4 | $^5/_2$ | $^9/_2$ |
| 8 | 3 | 4 | 4 |
| | 7 | $^8/_7$ | $^{48}/_7$ |
| 9 | 2 | 6 | 3 |
| | 6 | $^7/_3$ | $^{20}/_3$ |
| 10 | 5 | $^{18}/_5$ | $^{32}/_5$ |
| 11 | 4 | 5 | 6 |
| 12 | 3 | $^{20}/_3$ | $^{16}/_3$ |
| 13 | 2 | 9 | 4 |

Table 4. States arising in ionisation from $f^n$ free-ion configurations with first-order spin-orbit coupling. (Only multiplet components with normalised intensity ⩾ C.1 are listed)

| Initial state | | Final (LS) state and intensity | | Multiplet components ($J$) and intensities | |
|---|---|---|---|---|---|
| $f^2$ | $^3H_4$ | $^2F$ | 2.000 | $^5/_2$ | 1.714 |
| | | | | $^7/_2$ | 0.286 |
| $f^3$ | $^4I_{9/2}$ | $^3H$ | 2.333 | 4 | 1.890 |
| | | | | 5 | 0.424 |
| | | $^3F$ | 0.667 | 2 | 0.563 |
| $f^4$ | $^5I_4$ | $^4I$ | 2.545 | $^9/_2$ | 1.903 |
| | | | | $^{11}/_2$ | 0.599 |
| | | $^4G$ | 0.955 | $^5/_2$ | 0.658 |
| | | | | $^7/_2$ | 0.263 |
| | | $^4F$ | 0.500 | $^3/_2$ | 0.371 |
| | | | | $^5/_2$ | 0.114 |
| $f^5$ | $^6H_{5/2}$ | $^5I$ | 2.758 | 4 | 1.755 |
| | | | | 5 | 0.919 |

Table 4 (continued)

| Initial state | | Final (LS) state and intensity | | Multiplet components ($J$) and intensities | |
|---|---|---|---|---|---|
| | | $^5G$ | 1.266 | 2 | 0.513 |
| | | | | 3 | 0.575 |
| | | | | 4 | 0.165 |
| | | $^5F$ | 0.500 | 1 | 0.168 |
| | | | | 2 | 0.234 |
| | | $^5D$ | 0.476 | 0 | 0.149 |
| | | | | 1 | 0.224 |
| $f^6$ | $^7F_0$ | $^6H$ | 3.143 | $^5/_2$ | 0.898 |
| | | | | $^7/_2$ | 2.245 |
| | | $^6F$ | 2.000 | $^5/_2$ | 1.428 |
| | | | | $^7/_2$ | 0.571 |
| | | $^6P$ | 0.857 | $^5/_2$ | 0.816 |
| $f^7$ | $^8S_{7/2}$ | $^7F$ | 7.000 | 0 | 0.143 |
| | | | | 1 | 0.429 |
| | | | | 2 | 0.714 |
| | | | | 3 | 1.000 |
| | | | | 4 | 1.286 |
| | | | | 5 | 1.571 |
| | | | | 6 | 1.857 |
| $f^8$ | $^7F_6$ | $^8S$ | 1.143 | $^7/_2$ | 1.143 |
| | | $^6I$ | 1.857 | $^{15}/_2$ | 0.303 |
| | | | | $^{17}/_2$ | 1.500 |
| | | $^6H$ | 1.571 | $^{11}/_2$ | 0.101 |
| | | | | $^{13}/_2$ | 0.424 |
| | | | | $^{15}/_2$ | 1.030 |
| | | $^6G$ | 1.286 | $^9/_2$ | 0.134 |
| | | | | $^{11}/_2$ | 0.434 |
| | | | | $^{13}/_2$ | 0.694 |
| | | $^6F$ | 1.000 | $^7/_2$ | 0.138 |
| | | | | $^9/_2$ | 0.383 |
| | | | | $^{11}/_2$ | 0.460 |
| | | $^6D$ | 0.714 | $^5/_2$ | 0.107 |
| | | | | $^7/_2$ | 0.307 |
| | | | | $^9/_2$ | 0.301 |
| | | $^6P$ | 0.429 | $^5/_2$ | 0.230 |
| | | | | $^7/_2$ | 0.199 |
| $f^9$ | $^6H_{15/2}$ | $^7F$ | 2.333 | 5 | 0.583 |
| | | | | 6 | 1.674 |
| | | $^5L$ | 1.545 | 9 | 0.136 |
| | | | | 10 | 1.400 |
| | | $^5K$ | 1.364 | 8 | 0.211 |
| | | | | 9 | 1.131 |

Table 4 (continued)

| Initial state | Final (LS) state and intensity | | Multiplet components ($J$) and intensities | |
|---|---|---|---|---|
| | $^5I$ | 1.182* | 7 | 0.240 |
| | | | 8 | 0.913 |
| | $^5H$ | 1.000 | 6 | 0.233 |
| | | | 7 | 0.739 |
| | $^5G$ | 0.817* | 5 | 0.195 |
| | | | 6 | 0.605 |
| | $^5F$ | 0.310* | 5 | 0.232 |
| | $^5D$ | 0.454* | 4 | 0.454 |
| $f^{10}$ $^5I_8$ | $^6H$ | 2.800 | $^{11}/_2$ | 0.138 |
| | | | $^{13}/_2$ | 0.646 |
| | | | $^{15}/_2$ | 2.004 |
| | $^6F$ | 0.800 | $^{11}/_2$ | 0.727 |
| | $^4M$ | 1.462 | $^{21}/_2$ | 1.375 |
| | $^4L$ | 1.307 | $^{17}/_2$ | 0.134 |
| | | | $^{19}/_2$ | 1.167 |
| | $^4K$ | 1.189* | $^{15}/_2$ | 0.161 |
| | | | $^{17}/_2$ | 1.018 |
| | $^4I$ | 1.000* | $^{13}/_2$ | 0.154 |
| | | | $^{15}/_2$ | 0.837 |
| | $^4H$ | 0.379* | $^{13}/_2$ | 0.318 |
| | $^4G$ | 0.692* | $^{11}/_2$ | 0.611 |
| | $^4F$ | 0.405* | $^{9}/_2$ | 0.405 |
| $f^{11}$ $^4I_{15/2}$ | $^5I$ | 3.182 | 6 | 0.169 |
| | | | 7 | 0.736 |
| | | | 8 | 2.263 |
| | $^5G$ | 1.193 | 5 | 0.241 |
| | | | 6 | 0.913 |
| | $^5F$ | 0.625 | 5 | 0.550 |
| | $^3M$ | 1.462 | 10 | 1.400 |
| | $^3L$ | 1.307 | 9 | 1.206 |
| | $^3K$ | 1.154* | 7 | 0.117 |
| | | | 8 | 1.033 |
| | $^3I$ | 0.364* | 7 | 0.320 |
| | $^3H$ | 0.846* | 6 | 0.747 |
| | $^3G$ | 0.454* | 5 | 0.415 |
| | $^3F$ | 0.382* | 4 | 0.382 |
| $f^{12}$ $^3H_6$ | $^4I$ | 3.677 | $^{11}/_2$ | 0.212 |
| | | | $^{13}/_2$ | 0.914 |
| | | | $^{15}/_2$ | 2.536 |

Table 4 (continued)

| Initial state | Final (LS) state and intensity | | Multiplet components ($J$) and intensities | |
|---|---|---|---|---|
| | $^4G$ | 1.688 | $^7/_2$ | 0.106 |
| | | | $^9/_2$ | 0.419 |
| | | | $^{11}/_2$ | 1.157 |
| | $^4F$ | 0.667 | $^7/_2$ | 0.151 |
| | | | $^9/_2$ | 0.486 |
| | $^2L$ | 1.545 | $^{17}/_2$ | 1.500 |
| | $^2K$ | 1.364 | $^{15}/_2$ | 1.288 |
| | $^2I$ | 0.263 | $^{13}/_2$ | 0.242 |
| | $^2H$ | 1.000 | $^{11}/_2$ | 0.909 |
| | $^2G$ | 0.396* | $^9/_2$ | 0.359 |
| | $^2F$ | 0.470* | $^7/_2$ | 0.436 |
| | $^2D$ | 0.296* | $^5/_2$ | 0.296 |
| $f^{13}$ $^2F_{7/2}$ | $^3H$ | 4.714 | 4 | 0.321 |
| | | | 5 | 1.375 |
| | | | 6 | 3.018 |
| | $^3F$ | 3.000 | 2 | 0.357 |
| | | | 3 | 0.875 |
| | | | 4 | 1.768 |
| | $^3P$ | 1.286 | 0 | 0.107 |
| | | | 1 | 0.375 |
| | | | 2 | 0.804 |
| | $^1I$ | 1.857 | | |
| | $^1G$ | 1.286 | | |
| | $^1D$ | 0.714 | | |
| | $^1S$ | 0.143 | | |

Table 5. States arising from ionisation of $t_2^n$ configurations with spin-orbit coupling

| Initial state and spin-orbit component | | | Final states and spin-orbit components | | Intensities |
|---|---|---|---|---|---|
| $t_2^2$ | $^3T_1$ | $E$ or $T_2$ | $^2T_2$ | $U'$ | $^5/_3$ |
| | | | | $E''$ | $^1/_3$ |
| $t_2^3$ | $^4A_2$ | $U'$ | $^3T_1$ | $E$ | $^2/_3$ |
| | | | | $T_2$ | 1 |
| | | | | $T_1$ | 1 |
| | | | | $A_1$ | $^1/_3$ |
| $t_2^4$ | $^3T_1$ | $A_1$ | $^4A_2$ | $U'$ | $^4/_3$ |
| | | | $^2E$ | $U'$ | $^2/_3$ |
| | | | $^2T_1$ | $U'$ | 1 |
| | | | | $E'$ | 0 |
| | | | $^2T_2$ | $U'$ | $^1/_3$ |
| | | | | $E''$ | $^2/_3$ |
| $t_2^5$ | $^2T_2$ | $E''$ | $^3T_1$ | $A_1$ | $^2/_3$ |
| | | | | $T_1$ | $^3/_2$ |
| | | | | $E$ | $^1/_3$ |
| | | | | $T_2$ | $^1/_2$ |
| | | | $^1E$ | | $^2/_3$ |
| | | | $^1T_2$ | | 1 |
| | | | $^1A_1$ | | $^1/_3$ |
| $t_2^6$ | $^1A_1$ | $A_1$ | $^2T_2$ | $E''$ | 2 |
| | | | | $U'$ | 4 |

Table 6. States arising from ionisation of $t_2^n$ configurations in *j-j* coupling

| Initial configuration and state | | | Final configurations and states | | Intensities |
|---|---|---|---|---|---|
| $t_2^2$ | $(u')^2$ | $E$ or $T_2$ | $(u')^1$ | $U'$ | 2 |
| $t_2^3$ | $(u')^3$ | $U'$ | $(u')^2$ | $A_1$ | $^1/_2$ |
| | | | | $E$ | 1 |
| | | | | $T_2$ | $^3/_2$ |
| $t_2^4$ | $(u')^4$ | $A_1$ | $(u')^3$ | $U'$ | 4 |
| $t_2^5$ | $(u')^4(e'')$ | $E''$ | $(u')^4$ | $A_1$ | 1 |
| | | | $(u')^3(e'')$ | $T_1$ | $^3/_2$ |
| | | | | $E$ | 1 |
| | | | | $T_2$ | $^3/_2$ |
| $t_2^6$ | $(u')^4(e'')^2$ | $A_1$ | $(u')^4(e'')^1$ | $E''$ | 2 |
| | | | $(u')^3(e'')^2$ | $U'$ | 4 |

*Acknowledgements.* I would like to thank Dr. *A. F. Orchard* and Prof. *C. K. Jørgensen* for their interest and encouragement in this work. I am also grateful to the Royal Society for a European Exchange Programme Fellowship, and to Dr. *A. Veillard* for his hospitality at the Institut de Chimie in Strasbourg.

## 7. References

1. *Orchard, A. F.*: Electronic structure and magnetism of inorganic compounds. London: Chemical Society *1*, 1972; *2*, 1973; *3*, 1974.
2. *Evans, S., Hamnett, A., Orchard, A. F., Lloyd, D. R.*: Faraday Discussions Chem. Soc. *54*, 227 (1972).
3. *Evans, S., Green, M. L. H., Jewitt, B., King, G. H., Orchard, A. F.*: J.C.S. Faraday II *70*, 356, (1974).
4. *Rabelais, J. W., Werme, L. O., Bergmark, T., Karlsson, L., Husain, M., Siegbahn, K.*: J. Chem. Phys. *57*, 1185 (1972); *57*, 4508 (1972).
5. *Wertheim, G. K., Rosencwaig, A., Cohen, R. L., Guggenheim, H. J.*: Phys. Rev. Letters *27*, 505 (1971).
6. *Cox, P. A., Baer, Y., Jørgensen, C. K.*: Chem. Phys. Letters *22*, 433 (1973).
7. *Cox, P. A., Evans, S., Hamnett, A., Orchard, A. F.*: Chem. Phys. Letters *7*, 414 (1970).
8. *Fadley, C. S., Shirley, D. A.*: Phys. Rev. A *2*, 1109 (1970).
9. *Wertheim, G. K.*, and others: to be published.
10. *Cox, P. A., Orchard, A. F.*: Chem. Phys. Letters *7*, 273 (1970).
11. *Cox, P. A., Evans, S., Orchard, A. F.*: Chem. Phys. Letters *13*, 386 (1972).
12. *Cox, P. A.*: D. Phil Thesis. Oxford University 1973.
13. *Heitler, W.*: Quantum theory of Radiation. Oxford: Clarendon Press 1954.
14. *McWeeny, R., Sutcliffe, B. T.*: Methods of Molecular quantum mechanics. New York: Academic Press 1969.
15. *Condon, E. U., Shortley, G. H.*: Theory of atomic spectra. Cambridge: University Press 1935.
16. *Griffith, J. S.*: Theory of Transition metal ions. Cambridge: University Press 1964.
17. *Brink, D. M., Satchler, G. R.*: Angular momentum. Oxford: Clarendon Press 1968.
18. *Racah, G.*: Phys. Rev. *63*, 367 (1943).
19. *Racah, G.*: Phys. Rev. *76*, 1352 (1949).
20. *Nielson, C. W., Koster, G. F.*: Spectroscopic coefficients for $p^n$, $d^n$ and $f^n$ configurations. Cambridge, Mass.: M.I.T. Press 1963.
21. *Dixon, R. N., Hull, S. E.*: Chem. Phys. Letters *3*, 367 (1969).
22. *De Shalit, A., Talmi, I.*: Nuclear shell theory. New York: Academic Press 1963.
23. *Smith, A. L.*: Phil. Trans. A *268*, 169 (1970).
24. *Chang, T. N., Ishihara, T., Poe, R. T.*: Phys. Rev. Letters *27*, 838, (1971).
25. *Carlson, T. A., Krauze, M. O., Moddeman, W. E.*: J. Phys. (Colloques) *32*, C4 (1971)
26. *Potts, A. W., Williams, T. A.*: J. Electron Spectra Related Phenomena *3*, 3 (1974).
27. *Thülstrup, E. W., Ohrn, Y.*: J. Chem. Phys. *57*, 3716 (1972).
28. *Rohmer, M.-M., Veillard, A.*: Chem. Commun. *1973*, 251.
29. *Green, J. C., Green, M. L. H., Joachim, P. J., Orchard, A. F., Turner, D. W.*: Phil. Trans. A *268*, 111 (1970).

Received October 1, 1974

# X-Ray Photoelectron Spectroscopy
# Application to Metals and Alloys

**Richard E. Watson and Morris L. Perlman**

Departments of Physics and Chemistry, Brookhaven National Laboratory*, Upton, New York 11973 U.S.A.

**Table of Contents**



---

* Research performed under the auspices of the U.S. Energy Research and Development Administration.

# I. Introduction

In the mid-50's it was observed that the energy of a photoelectron, ejected from the core of an atom by an X-ray photon, is a rather sensitive probe of the chemical environment of the atom. From this observation has evolved a major research technique named electron spectroscopy for chemical analysis (ESCA) by the Uppsala group (*1*, *2*) which pioneered the subject and called X-ray photoemission spectroscopy (XPS) by many others. The field has developed rapidly; a third generation of spectrometers is in use at many laboratories and the understanding of the spectra observed is improving apace. A view of the current status of X-ray photoelectron spectroscopy in application to metals and alloys is presented in this article. We have not been encyclopedic in describing what has been done; we have instead attempted to cover the classes of results obtained and the kinds of problems encountered in interpretation of these results.

After a brief discussion of the thickness of the surface layer of a solid which is sampled by this technique, the various factors determining the position of a core level photoemission line are considered in some detail. The sudden approximation, for example, and the sum rule associated with it are examined; and data for two perhaps typical systems, Au—Sn and Ti—C, are analyzed. In both cases, the actual charge flow occurring between atomic sites in the alloy appears to be in a direction opposite to that indicated on first inspection of the spectra. The structures of the conduction band spectra of single elements and of alloys are then reviewed. Inasmuch as photoelectron energies are measured relative to the Fermi level, $\varepsilon_F$, the highest filled level in the conduction bands, chemical shifts associated with the bands are considered only in passing. Several other points are discussed in turn: the various sources of structure in core level spectra, 4*f* electron multiplet spectra (cf. *Jørgensen* in this volume), and surface states. Finally, there is a return to several matters of particular interest for discussion of the directions in which these may evolve. Consideration is given, for example, to the possible improvement of the crude but, we believe, effective analysis which is made of the core level shifts. Any review reflects to some extent the tastes of the authors; we hope the reader will find our choices useful.

# II. Escape Depth

At the outset of any discussion of X-ray photoemission from solids it should be noted that electron spectra can provide information about structure to a depth extending only a small distance from the emitting surface. X-rays may penetrate relatively far into the material but only those photoelectrons produced near the surface have any significant probability of escape without subsequent energy loss. Some representative data on escape depth, $\lambda$, as a function of electron energy, $E_0$, are shown in Fig. 1. Low energy results of *Kanter* (*3*) and of *Katrich* and *Sarbei* (*4*) for Au and higher energy results collected by *Powell* (*5*) for Cu, Ag, and Au are plotted. At the lowest energies the escape depth decreases with energy increase; this behavior depends in detail upon such factors as the plasmon excitation energy of the medium. In the vicinity of 10 to 100 eV, the depth is at a minimum

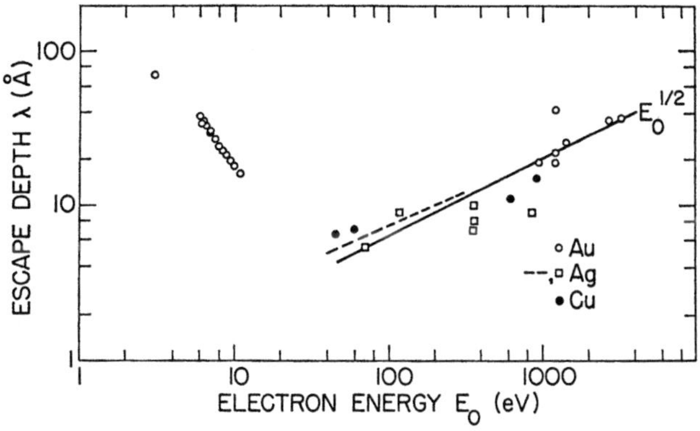


Fig. 1. Electron escape depths as a function of electron energy, $E_0$, in the noble metals. Ref. (3), (4), and (5)

of several Å, and the curve then rises gradually with a slope approximately that of the $E^{1/2}$ line on the figure. All the data available for the 50—3000 eV region, recently assembled by *Powell,* follow this universal curve if one allows the $\lambda$ values a scatter of about a factor of three. Powell's calculations suggest, however, that the scatter at a given electron energy would be greater if more classes of materials were represented. Among metals, for example, the alkalis would appear to have unusually large escape depths; no measurements for them have been made.

As *Powell* emphasizes, differences in experimental details and, in many cases, the lack of precise information about these details and about the specimen surfaces make the correlation of much of the existing escape data impracticable. What is important here is that at electron energies of 1253 and 1486 eV, corresponding to the Mg and Al $K_\alpha$ lines commonly employed in XPS, the escape depths are of the order of 20—40 Å. Photoexcitation of deeply bound core levels of course corresponds to smaller $E_0$ values and hence to smaller $\lambda$'s. As a rule an XPS experiment typically probes the outermost 15—20 Å of the sample. On one hand, the "bulk" is probed; on the other, XPS is a surface sensitive technique.

## III. Level Shifts

If the sample volume, from which photoelectrons are to be emitted, is put in electrical contact with the spectrometer, charge flows to establish the contact potential, which is the work function difference $\Phi_{\text{sample}} - \Phi_{\text{spect}}$, and the Fermi levels, $\varepsilon_F$, of the sample and spectrometer are equalized. This is shown schematically in Fig. 2. The work function in the experiment is $\Phi_{\text{spect}}$ and observation of the photoelectron kinetic energies from an atomic species in two different samples thus provides a measure of the shift of the level in question, level $i$, with respect to $\varepsilon_F$; if one wishes to consider the level shift with respect to some zero of the potential common to all the samples, then any shifts in $\varepsilon_F$ from sample to sample with respect to that same zero must be taken into account. This point

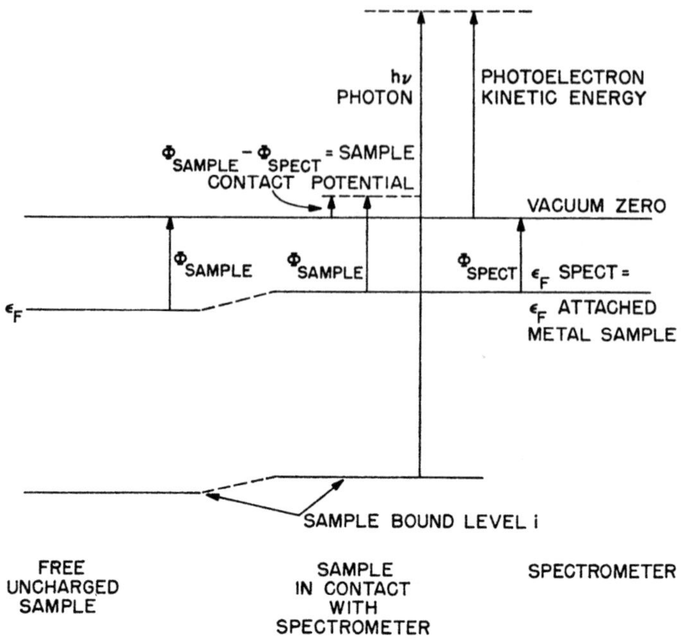


Fig. 2. Energy level scheme of the Fermi level, $\varepsilon_F$, and a bound level for a metallic sample. At left the sample is free and uncharged; at center it is in metallic contact with a photoelectron spectrometer and the contact potential has been established. The photoelectron kinetic energy associated with photoexcitation of the bound level is indicated

is to be pursued below. It is worth noting that shifts found in alloys and intermetallic compounds tend to be small, less than 1 eV, and it is necessary to guard against any spectrum instability caused by variation in the spectrometer work function (Fig. 2). A standard monitor line from a stable material, such as the $4f$ from Au, is often used for this purpose.

Let us, for the moment, invoke Koopmans' theorem (*6*) and associate the shift in a photoemission line with a shift of the appropriate one-electron energy, $\varepsilon_i$. If the electron orbital, $\phi_i$, is sufficiently corelike so as not to be modified spatially by alteration of the atom's environment, then the change in electron density $\Delta\varrho_e$ and the nuclear charge difference $\Delta\varrho_N$ in the region surrounding $\phi_i$ lead to an energy shift as given by the Hartree-Fock approximation,

$$\Delta\varepsilon_i = \int \left|\phi_i(r_1)\right|^2 \frac{1}{r_{12}} [\Delta\varrho_e(r_2) + \Delta\varrho_N(r_2)] d\tau_2 + \text{exchange terms associated with } \Delta\varrho_e. \quad (1)$$

The reference in this expression is implicitly the vacuum zero if the sample had no surface "dipole" barrier potential $\phi_b$ (Fig. 3a). In order to relate a measured photoelectron energy shift, which is referred to individual Fermi levels, to such a $\Delta\varepsilon_i$, we must include an estimate of any Fermi level shift with respect to *this* crystal vacuum zero. An experimental value for the work function, $\Phi$, is, un-

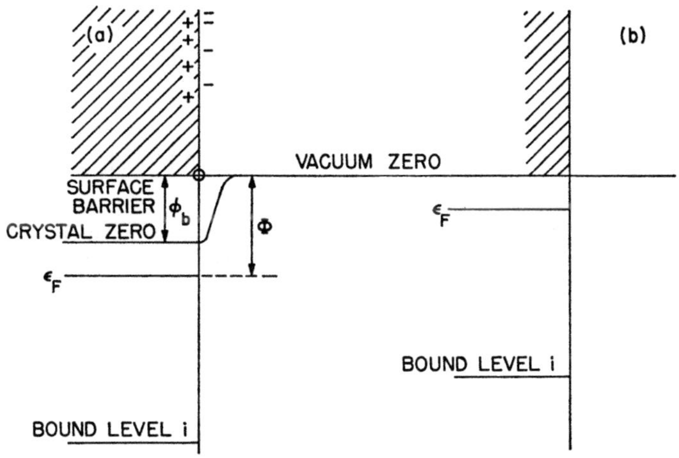


Fig. 3. The scheme of a bound level and $\varepsilon_F$ for a free uncharged sample a) with its surface dipole barrier term, $\phi_b$, and b) with the dipole barrier set to zero. Equations for the chemical shift in a bound level of a solid, as given in this review and in most of the published literature, do not include $\phi_b$. Behavior of $\phi_b$ must be taken into account if experimental work functions, $\Phi$, are utilized to estimate the chemical shift of $\varepsilon_F$

fortunately, the difference between the ordinary vacuum zero and the Fermi level; *i.e.*, it includes $\phi_b$. A binding energy shift, $\Delta\mathscr{E}$, is thus related to $\Delta\varepsilon_i$ by

$$\begin{aligned} -\Delta\mathscr{E} &= \Delta\varepsilon_i + \Delta(\Phi - \phi_b) \\ &\equiv \Delta\varepsilon_i - \Delta\varepsilon_F \end{aligned} \tag{2}$$

Following convention, $\mathscr{E}$ has here been taken as a positive quantity, whereas $\varepsilon_i$ and $\varepsilon_F$, measured with respect to the vacuum zero, are negative. There has been some success in the calculation of $\Phi$ values for simple metals (7); this is much less the case, however, for either the barrier term or for $(\Phi - \phi_b)$ separately. There has been a tendency in the literature to employ experimentally determined $\Phi$ values and replace $\Delta\varepsilon_F$ by $-\Delta\Phi$. This usually unavoidable approximation taken with the scatter often encountered in experimental $\Phi$ values, may introduce an error of as much as several eV into the analysis of a chemical shift. In attempting to bypass this problem by comparing differences of shifts between two levels, one introduces other problems, as will be discussed in Section IV—B.

Both the initial $N$-electron and the final $(N-1)$-electron states involved in a photoemission event are proper quantum mechanical states; thus for a free atom the photoelectron binding energy is

$$+\mathscr{E} = E_F^{\text{exact}}(N-1) - E_I^{\text{exact}}(N), \tag{3}$$

where $E_F$ and $E_I$ are the exact final and initial state energies, respectively. It is convenient to rewrite this expression in terms of Hartree-Fock total energies, which are related to the exact by

$$E^{\text{exact}} \equiv E^{\text{H-F}} + E^{\text{C}}.$$

Here $E^{\mathrm{C}}$ is the correlation energy of the state in question. At least for the lowest lying state of some given symmetry, $E^{\mathrm{H-F}}$ lies higher in energy than $E^{\mathrm{exact}}$ and $E^{\mathrm{C}}$ is negative. Eq. (3) becomes

$$\mathscr{E} = E_{\mathrm{F}}^{\mathrm{H-F}}(N-1) - E_{\mathrm{I}}^{\mathrm{H-F}}(N) + \Delta_{\mathrm{C}} . \tag{4}$$

In general, correlation effects are greater in the system with the greater number of electrons, thus $E_{\mathrm{I}}^{\mathrm{C}} < E_{\mathrm{F}}^{\mathrm{C}}$ and $\Delta_{\mathrm{C}}$ is positive. In terms of the Hartree-Fock energies, the Koopmans' theorem based Eq. (1) is equivalent to the statement

$$-\varepsilon_i = E_{\mathrm{F}}^{\mathrm{H-F}}(N-1) - E_{\mathrm{I}}^{\mathrm{H-F}}(N),$$

and, in the absence of correlation effects, it could hold only if the $(N-1)$ electrons in the final system did not relax about the hole left by electron $i$. In reality, relaxation acts to lower the final state energy, and Eq. (4) for the free atom becomes

$$\mathscr{E} = -\varepsilon_i + \Delta_{\mathrm{C}} + \Delta_{\mathrm{R}} , \tag{5a}$$

or for the solid, in which the binding energy is measured relative to the Fermi level,

$$\mathscr{E} - \varepsilon_{\mathrm{F}} = -\varepsilon_i + \Delta_{\mathrm{C}} + \Delta_{\mathrm{R}} . \tag{5b}$$

In Eq. (5) the relaxation term, $\Delta_{\mathrm{R}}$, is negative and $\Delta_{\mathrm{C}}$, as already mentioned, is positive. This cancellation property has the fortuitous result that the ionization energy of a valence electron in a free atom tends to be given more accurately by $\varepsilon_i$ than by the difference in Hartree-Fock total energies. Of course, for deeper lying electrons relaxation is more important; the hole simulates a change in nuclear change of $\simeq +1$ to any electrons lying outside it and $\Delta_{\mathrm{R}}$ becomes substantially larger than $\Delta_{\mathrm{C}}$ in such a case. *Hedin* and *Johansson* have inspected (*8*) this trend, shell by shell, for Na and K.

## A. The Sudden Approximation and the Sum Rule

Even if relaxation effects make the binding energy depart appreciably from $\varepsilon_i$, Eq. (1) can still be useful in the quantitative analysis of a line shift: in a discussion of conduction electron relaxation effects in metals *Lundqvist* stated (*9*) "The polarization of the conduction electrons displaces the core quasiparticle level by a considerable amount and introduces a characteristic satellite structure into the spectrum, but it does not affect the average energy of the core spectrum". In other words there is a sum rule acting so as directly to relate $\varepsilon_i$ to the center of gravity of the ith electron spectrum. Measurement of this spectrum in sufficient detail to determine its center of gravity is generally not practicable. Nevertheless the sum rule does offer the possibility of relating the shift of the principal line of a spectrum to a shift in $\varepsilon_i$, and hence to Eqs. (1) and (2), whenever the several environments, into which the sampling atom is placed, display similar polarization responses. In such cases, the use of the relation

$$-\Delta\mathscr{E}' = \Delta\varepsilon_i - \Delta\varepsilon_F \, , \tag{6}$$

in which $\mathscr{E}'$ is the apparent binding energy associated with some principle feature of the spectrum, greatly simplifies the problem of relating a line shift to the change in charge distribution at an atomic site and its vicinity. A derivation of the sum rule is sketched below. Detailed discussions on the sum rule as applied to atoms (*10*) and metals (*9, 11*) have been published.

The basic assumption in the derivation, one generally considered valid for the time scale involved in photoemission, is the "sudden approximation": that is, in the wave function, $\Phi_i(N-1)$, for the $(N-1)$ electron ion immediately after photoemission, all the $(N-1)$ orbitals, except of course that of the missing ith electron, are frozen in the forms they had in the initial $N$ electron state $\Psi_0(N)$. Now this $\Phi_i$ is not an eigenfunction of the actual $(N-1)$ electron system, but it may be expanded in a complete set of such eigenfunctions, $\Psi_n(N-1)$,

$$\Phi_i(N-1) = \sum_n [\langle \Phi_i | \Psi_n \rangle \Psi_n(N-1)] \, , \tag{7}$$

where $\langle \Phi_i | \Psi_n \rangle$ is an overlap integral. The probability, $P_{in}$, of being in the nth state after photoemission is simply

$$P_{in} = |\langle \Phi_i | \Psi_n \rangle|^2 \, . \tag{8}$$

Now the binding energy corresponding to the center of gravity of the ith electron photoemission spectrum is

$$\begin{aligned} \bar{E}_i - E_0 = & \sum_n [P_{in} \int \Psi_n^*(N-1) H \Psi_n(N-1) d\tau_1 \ldots d\tau_{N-1}] \\ & - \int \Psi_0^*(N) H \Psi_0(N) d\tau_1 \ldots d\tau_N \, , \end{aligned} \tag{9a}$$

where $H$ is the many-electron Hamiltonian. Taking into account the fact that the $\Psi_n$'s are eigenfunctions of $H$, that is

$$\int \Psi_m^*(N-1) H \Psi_l(N-1) d\tau_1 \ldots d\tau_{N-1} = E_l \delta_{m,l} \, ,$$

and substituting Eqs. (7) and (8) into (9a), one has

$$\begin{aligned} \bar{E}_i - E_0 = & \int \Phi_i^*(N-1) H \Phi_i(N-1) d\tau_1 \ldots d\tau_{N-1} \\ & - \int \Psi_0^*(N) H \Psi_0(N) d\tau_1 \ldots d\tau_N \, . \end{aligned} \tag{9b}$$

If the initial state function $\Psi_0$, and hence $\Phi_i$ as well, is a single determinant, the right hand side is $-\varepsilon_i$ by definition. Thus

$$\bar{E}_i - E_0 = -\varepsilon_i \, , \tag{9c}$$

or in other words the center of gravity of the spectrum corresponds to the one-electron energy. It may be worth noting again that this sum rule rests on two important assumptions; first, that the sudden approximation is a valid description of the photoionization process (this is generally accepted); and second, that, in the initial state, correlation effects involving the electron $i$ are insignificant (see note added in proof).

## B. Atom to Solid Metal Level Shifts

As a prelude to considering chemical shifts associated with alloying, let us first consider the measured level energy shift on going from a free metal atom to the solid metal. First, there is the fact that the energy for the free atom is referred to the vacuum zero, while for the solid the reference is generally the Fermi level. A shift is thus introduced and, as discussed earlier, there is no salvation in the work function. Second, electron configuration often changes on going from the free atom to metal; for example, transition metal atoms tend to have $d^n s^2$ ground states whereas the solid tends to approximate a $d^{n+1}s^1$ configuration. Third, there is a substantial compression, or "renormalization", of the valence electron charge on forming the metal, and this, in the sense of Eq. (1), makes an important contribution to the level shift. Fourth, in respect to a valence level, there are, of course, shifts and structure associated with broadening of the level into a conduction band. Fifth, relaxation effects differ; photoemission from a neutral atom leaves behind an ion of charge $+1$, whereas, in metal, conduction electrons and cores of neighboring atoms tend to screen the vacancy. Finally, there may be contributions from correlation energy differences. These factors are considered in order below.

Level positions of a metal sample are measured with respect to its Fermi level as already discussed with reference to Figs. 2 and 3. Thus, with respect to the free atom, there is introduced, in effect, a level shift equal to the difference between $\varepsilon_F$ and the crystal zero, the latter corresponding to vacuum zero for a free atom. This shift reduces the apparent binding energy usually by an eV or thereabouts, an amount less than the work function.

A classic example of change of electron configuration on forming a solid is seen in diamond; the atomic $2s^2 2p^2$ configuration goes over to the $2s2p^3$ necessary for the formation of tetrahedral bonds. The energy required, 4 eV per atom, is more than made up by bonding energy. Again, in transition metals, when the bands are filled to $\varepsilon_F$ the electrons are described as being approximately in a $d^{n+1}s^1$ configuration, where "$s$" is taken to be any non-$d$. There is a significant shift to be associated with the difference between this metal configuration and that for the free atom, $d^n s^2$. In fact, for atomic $3d^8 4s^2$Ni the spherical Coulomb interactions of a $2s$ electron with single $3d$ and $4s$ electrons are the Slater integrals *(12)*

$$F^\circ(3d, 2s) = 37 \text{ eV and}$$

$$F^\circ(4s, 2s) = 11 \text{ eV},$$

where

$$F^\circ(i, n) \equiv \iint P_i^2(r_1) \frac{1}{r_>} P_j^2(r_2) r_1^2 dr_1 r_2^2 dr_2 ;$$

$r_>$ is the greater of the two radial coordinates $r_1$ and $r_2$, and the $P(r)$ are radial components of the one-electron orbitals $i$ and $j$. Thus, if there were no modifications, *i. e.* relaxation, of the valence electron orbitals themselves concomitant with the $s$—$d$ interchange in the atom → metal transition, the $2s$ level energy would undergo a Coulomb shift equal to the difference in the Slater integrals, +26 eV. Valence orbital modification reduces this to +5 1/2 eV, still a substantial shift.

If the Wigner-Seitz cell appropriate to a metal is superimposed on the spatial charge distribution of a free atom, one finds characteristically that a quantity of charge, typically between 2/3 e and 1 e, lies outside the cell boundaries. (*13*) Since in the metal the cell is of course neutral, this implies that formation of the metal requires compression of the valence charge, and associated with the compression is an increase of the Coulomb interactions of the valence electrons with each other and with the ion core. A lowest order estimate of the shift associated with this effect may be based on truncation of the free atom valence orbitals at the cell radius, $r_{WS}$, and renormalization of the charge within the cell. For a core electron lying entirely inside the valence density the core-valence Coulomb interaction is

$$F^{\circ}(\text{core, val}) = 27\left\langle \frac{1}{r} \right\rangle_{\text{val}} \text{eV},$$

where $\langle 1/r \rangle$, the expectation value of the valence electrons, is given in Bohr units, (au). In the free atom, that part of the charge lying outside the Wigner-Seitz cell is centered, typically, at 5—7 au; in the metal this charge is located at considerably smaller radius, in the vicinity of 3 au. This compression, if a charge 1e is involved, leads to a core level shift roughly

$$\Delta\varepsilon_i \simeq \frac{27}{3} - \frac{27}{6} \simeq 4 \text{ to } 5 \text{ eV}.$$

Taken together then, in metals the effects of valence charge compression and possible change in valence configuration raise core level $\varepsilon_i$ values so that they are 5—9 eV higher than they are in free atoms. Although this may seem to be an embarrassingly large figure in view of the fact that the cohesive energies of metals are only 1 to 8 eV per atom, there is in actuality no difficulty; the total Hartree-Fock energy of an $N$ electron system is

$$E_{\text{tot}} = \sum_i \varepsilon_i - [\text{doubly counted interelectronic terms}] \tag{10}$$

and the increase in the $\varepsilon_i$ sum is approximately cancelled by increase in the second term. This is nicely illustrated in the case of the alkali metals. Were it not for the requirement of a kinetic energy increase in excess of the potential energy decrease, the single valence electron in the free alkali atom would contract into the attractive positive ion core region. In the solid, however, the boundary conditions are different (*13, 14*); for example, the state at the bottom of the bands, $\Gamma_1$, has a wave function which is flat at $r_{ws}$ (Fig. 4), and the electron may therefore move

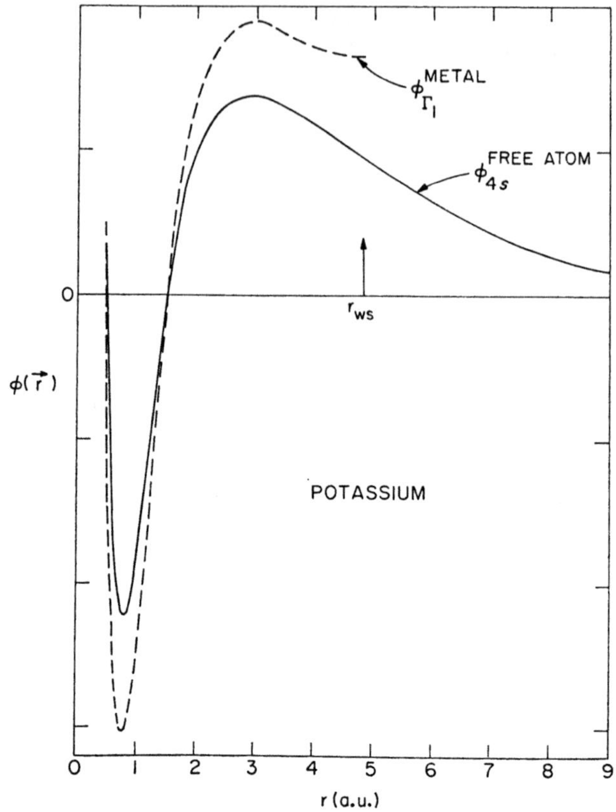


Fig. 4. Hartree-Fock free atom 4*s* valence electron orbital for potassium (solid line) and the 4*s*-like orbital, obeying the Wigner-Seitz boundary condition, appropriate to the bottom of the conduction bands in metallic potassium (dashed line). Both orbitals are normalized, $\int |\phi|^2 dr = 1$; for the metal, integration is limited to the Wigner-Seitz sphere of radius $r_{WS}$

in to smaller average radius without requirement of excessive kinetic energy[1]). It is this contraction, allowed by the boundary conditions in the solid, which produces the "bonding" wave function shown in Fig. 4 and the cohesive energy of the alkali metals (*14*). If the metal is assumed to have free electron bands, the average energy of the occupied levels in the band is

$$\varepsilon_{av} = \varepsilon_{\Gamma_1} + \frac{3}{5} k_F^2 ,$$

where $k_F$ is the wave vector at the Fermi level[2]). Since there is just one valence electron per atom, the cohesive energy,

---

[1]) Although the virial theorem has been ignored in this argument and in some which follow, the theorem is nonetheless satisfied in the conclusions drawn.

[2]) If the band is free electron like, $k_F$ is simply a function of the lattice constant and the number of electrons per atom in the band.

$$E_{\text{coh}} = E_{\text{tot}}(\text{metal}) - E_{\text{tot}}(\text{free atom}) ,$$

is obtained simply by use of Eq. (10). Thus substituting, one has

$$E_{\text{coh}} = (E_{\text{core}} + \varepsilon_{\Gamma 1} + \frac{3}{5} k_{\text{F}}^2)_{\text{metal}} - (E_{\text{core}} + \varepsilon_{\text{val}})_{\text{free atom}} ;$$

here $E_{\text{core}}$ is the bare core energy, which does not include valence-core interelectronic terms. Assuming the same core in the metal as in the free atom, we have

$$E_{\text{coh}} = \varepsilon_{\Gamma_1} + \frac{3}{5} k_{\text{F}}^2 - \varepsilon_{\text{val, free atom}} . \tag{11}$$

With Hartree-Fock estimates for $\varepsilon_{\Gamma_1}$ and for the free atom valence $s$ electron energy, Eq. (11) reproduces the measured cohesive energy (*15*) within $\sim 10\%$. Valence electron-core electron interaction of course increases in the condensation process, and the level for each core electron, referenced to the vacuum zero, lies shallower in the metal than in the atom by something like $E_{\text{coh}}$. If this would seem implausible, it is only because the second term of Eq. (10) has not properly been taken into account.

For polyvalent and transition metals the situation, though more complicated, is similar to that outlined for the alkalis. When more than one valence electron is involved, the valence-valence Coulomb contributions to the second term of Eq. (10) result in relaxation of the requirement that the center of gravity of the occupied band for the metal, with no surface barrier $\phi_{\text{b}}$, should lie lower than the free atom valence level. In the transition metal atoms, the $d$ orbitals are corelike and undergo little compression when the $d$ band forms on condensation to solid; and the center of gravity of the $d$ band, $\varepsilon_d$, behaves rather like a core level, moving upward due mostly to compression of the non-$d$ valence electrons and with lesser effects produced by compression of each of the $n-1$ other $d$ electrons. The variation in cohesive energy over a transition metal series is directly associated with the width and occupation of the bands (*16*). At the light end of a series the low lying "bonding" levels are filling, but in the bands of the heavier members the higher lying "non-bonding" levels are increasingly occupied (*17*); thus, the relative energy stabilization is less and, as a result, $E_{\text{coh}}$ is largest in the middle of the series. For any quantitative estimate of $E_{\text{coh}}$ valence-valence terms and the exact position of $\varepsilon_d$ must properly be taken into account.

One has an intuitive, simple description of transition metal behavior in terms of properties which depend on the atomic potential, such as $\varepsilon_d$, and other properties, such as the $d$ band structure, which depend on crystal symmetry and lattice constant. This division is useful for consideration of XPS conduction band results in transition metal alloys and it may be noted that core and $d$ levels, alike, lie higher in the metal than in the free atom.

A free atom is left with a charge $+1$ after photoelectron emission; in a metal, however, conduction electrons act to screen the vacancy. This screening may be thought of as "complete" if the Wigner-Seitz cell containing the hole is essentially at charge neutrality. Experimental and theoretical studies of the extent of screening are being done. For $4f$ excitations in the rare earth metals, *Herbst et al.* (*18*) found that the one-electron energies, $\varepsilon_{4f}$, computed for these metals, though higher (shallower) than the free-atom values, are significantly lower (deeper) than the measured excitation energies. Relaxation effects may be important, not surprising in view of the fact that the $4f$ electrons are in the core region; the effects of intra-atomic and conduction electron screening were estimated in the complete screening limit by calculating

$$E_{\text{excited}}(\ldots .\ 4f^{n-1}5s^2 5p^6 6s5d^{x+1}) - E_{\text{ground}}(\ldots .\ 4f^n 5s^2 5p^6 6s5d^x)\ . \quad (12)$$

Ground state $6s$ and $5d$ counts were chosen consistent with band theory arguments; relativistic and correlation effects were included by using in Eq. (12) values based on optical data for free atoms with corrections, obtained from Hartree-Fock calculations, for the difference between free atom and band structured metal. The results are plotted in Fig. 5. Agreement with experiment is extraordinarily good. By way of comparison, relativistic estimates of one-electron energies typically place the levels 3 to 4 eV deeper. Data are now becoming available for other rare-earth levels: for $5p$, in contrast with the $4f$ case, relativistic estimates for the one-electron energies in the metals predict the ionization energies to within $\sim$1 eV. The fact that the one-electron energies do not fit $4f$ ionization energies of the rare earths and do fit $5p$, the outermost core shell, has been interpreted, perhaps simplistically, to mean that screening is relatively unimportant in outer core ionization. There remain open questions about screening by conduction electrons.

*Ley* and co-workers (*19*) have treated conduction electron screening of core holes for the $3d$ and other metals by a method which in a way approximates the

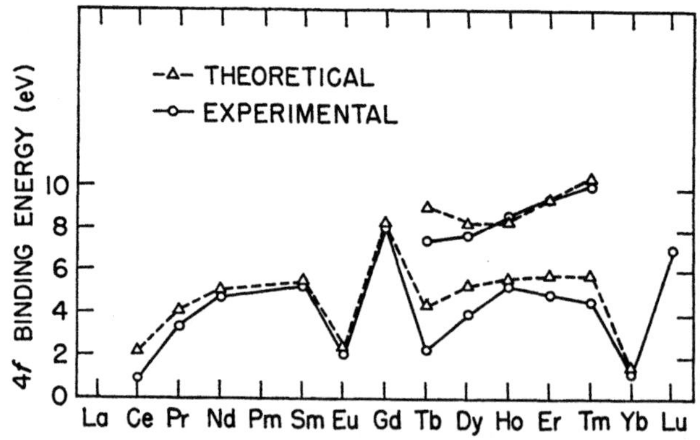


Fig. 5. Experimental [Ref. (*36*)] and theoretical [Ref. (*18*)] $4f^n \rightarrow 4f^{n-1}$ photoexcitation energies for the rare-earth metals. For the theoretical estimates "complete" screening of the $4f$ vacancy was assumed (see text). The lower curves are for photoexcitation into the lowest multiplet level of the $4f^{n-1}$ system. When the initial state involves an over half-filled ($n > 7$) $4f$ shell of spin $S$, the final states may have spins $S + 1/2$ *and* $S - 1/2$. The lower curves for Tb through Tm are for spin $S + 1/2$; the upper curves are for excitation to spin $S - 1/2$

total energy difference. Following an idea previously applied to atoms, (*20*) they calculated the one-electron core level energy in the average of the initial and final state potentials. In this scheme the net effect of any screening by extra-atomic conduction electrons in the final state reduces to half the Coulomb and exchange interaction of the screening charge with the core orbital. They further assumed that the screening charge is in a single partial wave having the lowest quantum numbers, $n'l'$, consistent with the orbitals already occupied at the atomic site. Thus, the screening contribution to the binding energy becomes

$$\Delta = -\frac{1}{2}\,[F^{\circ}(n'l', \text{core}) - \text{exchange terms}]_{Z+1}\,, \qquad (13)$$

where the reference is to $Z+1$ because, for a screening electron, an atom of atomic number $Z$ with a deep core vacancy is equivalent to an atom with nuclear charge $Z+1$. For Sc through Ni, $n'l'$ was taken to be $3d$; the possible effect of partial occupation of the $3d$ bands on screening was not considered. Almost all of the charge associated with the $3d$ function in the transition metals lies within the W—S cell; thus these cases are essentially in the "complete" screening limit. The $\Delta$'s resulting from this approach range from $-13$ eV for Ti to $-18$ eV for Ni, values too large by a factor at least three. Results are much better for metals such as Na, Al, Mg, and Cd, in which the screening charge partial wave functions are more diffuse.

Finally, one should not overlook the possible role of correlation effects in atom-metal differences. In atoms the dominant contributions to the correlation energy arise from interaction between electrons of the same principal quantum number $n$, since these have the greatest overlap. As far as purely intraatomic electrons are concerned, the correlation terms differ very little between atom and metal; correlation does, of course, affect interatomic screening of the final state vacancy in metals.

*Ley* and co-workers (*19*) have presented quasi-experimental free atom-to-metal core level shifts for a sequence of elements, Ti through Zn. Free atom photoemission data do not exist in most of these cases, and *Ley et al.* used, instead, ground state calculations in which the neutral atoms were given $3d^n4s^2$ configurations except for Cr ($d^5s$) and Cu ($d^{10}s$). Their core level binding energy shifts,

$$\Delta\mathscr{E} = [\mathscr{E}\,(\text{measured, metal}) + \Phi] - \mathscr{E}\,(\text{calculated, free atom}), \qquad (14)$$

are shown as the filled circles in Fig. 6. Measured binding energies for metals were increased by the amount of the work function to take into account the fact that these energies are determined with respect to $\varepsilon_F$ rather than to a vacuum or crystal zero. As already discussed, such a correction, though convenient, actually does not provide the correct referencing to the crystal zero, namely $\Phi - \phi_b$. Subtraction of the unwanted barrier contribution would make $\Delta\mathscr{E}$, a negative quantity, 1—2 eV more negative than the quantities given in Fig. 6. The *Ley et al.* estimates of $\Delta\mathscr{E}$, solid circles, show a notable break between Ni and Cu. This the investigators ascribed to the disappearance of $3d$ screening, no longer possible

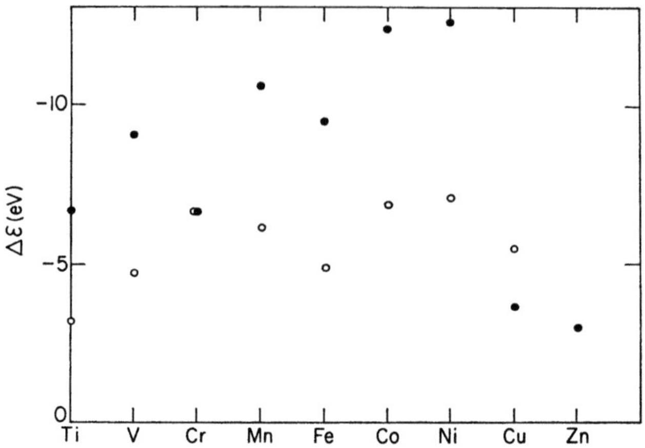


Fig. 6. $\Delta\mathcal{E}$, the difference between free atom and metallic binding energies of $2s$ or $2p$ core levels in the elements Ti through Zn. Estimates by *Ley et al.* [Ref. (*19*)], shown as filled circles, are based on $d^n s^m$ configurations which were taken to differ between atom and metal for all elements except Cr and Zn. The open circles are values based on free atom binding energies for atom configurations $d^\nu s^\mu$ most closely corresponding to the metallic $d$ electron counts. (See text.)

once the $3d$ bands are filled, as they were assumed to be in Cu. An alternative explanation for this behavior is suggested by the deviation of the point for Cr. One of the quantities that affects $\Delta\mathcal{E}$ is the difference in electron configuration between that of the free atom and that in metal. Any abrupt change in this configuration difference would by itself produce a break in the $\Delta\mathcal{E}$ trend. It is thus useful to calculate $\Delta\mathcal{E}$ with configurations which are taken to be the same for the free atom as they are for metal. Now general band theory experience indicates that the metals Ti through Ni are well described by $3d^{n+1}4s^1$ configurations. In Cu, however, though the $d$ bands are filled, hybridization with conduction electron states *above* $\varepsilon_F$ depletes the count of $d$ character in the occupied bands; Cu thus deviates from the $d^{n+1}s^1$ description. Band theory results indicate an approximate $d^{9.75}s^{1.25}$ assignment (*13, 21*). Corrected values of $\Delta\mathcal{E}$ have been calculated for free atom configurations corresponding to those for the metals by use of differences in one-electron energies obtained from free-atom Hartree-Fock calculations for $3d^{n+1}4s^1$ and $3d^n4s^2$ configurations. In the case of Cu the $d^{9.75}s^{1.25}$ free atom value was obtained by interpolation. These corrected $\Delta\mathcal{E}$ points are shown in Fig. 6 as open circles. For Cr the two points coincide because *Ley et al.* indeed used the $d^{n+1}s^1$ free atom configuration. The sharp break between Ni and Cu is no longer in evidence and the corrected $\Delta\mathcal{E}$ values range from —3 to —7 eV. This energy arises from extraatomic relaxation, from charge renormalization and from any accumulated errors such as the omission of $\phi_b$. Self-consistent band calculations, made for all the metals except Zn, provide estimates of the renormalization shifts, $\sim -3$ eV for Ti and Cu and $-4$ to $-5$ eV for the other metals (*21*). These values suggest quite modest contributions to $\Delta\mathcal{E}$ from extraatomic relaxation.

## IV. Core Level Shifts with Alloying

When an atom is taken from one metallic environment to another, the electron binding energy shifts tend to be less than 1 eV. Shifts are small because, for one thing, ionicity is small in metals; that is, any tendency toward substantial charging is screened by the conduction electrons. Further, any significant chemical shift in a core level energy is associated with a similar shift in $\varepsilon_F$, and the binding energy, normally measured with respect to $\varepsilon_F$ [Eq. (2)], therefore changes little. This is particularly true when $\varepsilon_F$ is located within a band having a high density of states, as is the case for transition metals such as Ni, with its *d* bands; a shift of even 0.1 eV in the position of $\varepsilon_F$ in the band would require an unsupportably large change of total charge in each atomic cell. Unless alloying is associated with substantial modification of the width or structure of the *d* bands, $\varepsilon_F$ is thus pinned with respect to the *d* band center of gravity, $\varepsilon_d$; and Eq. (2) may be rewritten

$$-\Delta\mathscr{E} \sim \Delta\varepsilon_i - \Delta\varepsilon_d\,. \tag{2a}$$

Because, in first approximation, $\varepsilon_d$ and a core level energy $\varepsilon_i$ undergo much the same chemical shift, $\Delta\mathscr{E}$ tends to be quite small.

This being the situation, it is imperative to choose for measurement the narrowest accessible core lines, as a rule those from the levels of highest *n* and *l*. For the 3*d* and 5*d* transition metals the 2*p* and 4*f* lines respectively are best.

To the extent that extraatomic relaxation is unaffected by change of environment, the shift in a photoemission line provides a measure of the one-electron energy shift in the sense of Eqs. (1) and (2). Extraatomic relaxation should be more nearly the same between two metals than between, say, metal and insulator. Rather than assuming constancy of relaxation, one might try to measure not just the main line originating from a given level but all the satellites as well and invoke the sudden approximation sum rule [Eq. (9b)] to arrive at $\Delta\varepsilon_i$. Unfortunately, much of this detailed structure is unresolved and usually hidden in the background at electron kinetic energies lower than those of the principal lines. One is perforce satisfied to monitor, for any evidence of environment-dependent relaxation, either the principal satellites, which in metals arise from plasmon excitations or, in many cases, the overall shape of the line and nearby background.

Core level shifts in two alloy systems Au—Sn and Ti—C, will be considered as examples in the remainder of this section. Gold, the most electronegative of the elements, is of interest in itself. In Au—Sn volume is conserved between the alloy and its pure consituents, *i.e.*,

$$V(\mathrm{Au}_x\,\mathrm{Sn}_y) \simeq xV(\mathrm{Au\ metal}) + yV(\beta-\mathrm{Sn})$$

where the $V$'s are atomic volumes. Moreover, there are available Mössbauer isomer shift data, which may be usefully combined with the photoemission results. The second example, TiC, is a transition metal compound with volume effects. Here, *Ramqvist* and co-workers (22) have endeavored to circumvent the problem of Fermi level shift by studying relative shifts of different core levels in a given atom. As will be seen, this approach has its own difficulties.

A. The $Au_xSn_y$ System

Measured core level shifts are listed in Table 1 for three Au—Sn alloys *(23)*, two of which are ordered. Negative values indicate that the levels are more tightly bound in the alloys, and the rather substantial negative value for Au in $AuSn_4$

Table 1. Core-level shifts of Au and Sn in the $Au_xSn_y$ system[a])

| Sample | Level-energy shift | |
|---|---|---|
| | $\Delta_{4f}$(Au) (eV) | $\Delta_{3d}$(Sn) (eV) |
| Au | 0 | |
| $Au^{0.96}Sn^{0.04}$ | $-0.10\pm0.14$ | $-0.40\pm0.14$ |
| AuSn | $-0.45\pm0.14$ | $-0.4\ \pm0.14$ |
| $AuSn_4$ | $-1.10\pm0.14$ | $-0.1\ \pm0.14$ |
| Sn | | 0 |

[a]) Ref. *(23)*.

suggests, in the sense of Eq. (1), that charge flows from Au to Sn sites. Mössbauer isomer shifts, too, depend upon differences in conduction electron configuration according to the relation[24]

$$IS = \frac{4\pi Ze^2R^2c}{5E\gamma}\,\frac{\delta R}{R}\,\Delta\varrho\,(0)\,, \tag{15}$$

where $E_\gamma$ is the energy of the nuclear transition, $\delta R/R$ is the fractional change in nuclear charge radius on deexcitation, and $\Delta\varrho(0)$ is the difference in "contact" electron density at the nucleus between the two samples being compared, the Mössbauer source and the Mössbauer absorber. If the nuclear parameters are known, then the Mössbauer shift provides a measure of the change in the amount of *s*-like (and relativistic $p_{1/2}$-like) charge density in the atomic cell surrounding the nucleus in question. Data for both Au and Sn sites in the Au—Sn system are collected in Table 2 *(25)*. These data together with the nuclear terms[3]) show that $\varrho(0)$ for Au increases substantially upon alloying. This behavior, typical for Au, has been viewed as indicating substantial conduction charge flow onto Au sites, a view consistent with the large electronegativity assigned Au but opposite to the trend suggested by the level energy shifts. The latter, of course, reflect all changes in conduction electron distribution via the Coulomb and exchange interactions

[3]) Conversion of an isomer shift to a change in electron count of course requires values for the nuclear parameters, which include the factor $\Delta R/R$, the fractional change in nuclear charge radius on excitation in the Mössbauer transition. For $^{197}$Au the value $\Delta R/R = +3\times10^{-4}$ was employed [ref. *(24)*]. In the case of $^{119}$Sn the value chosen, $(+1.8\pm0.4)\times10^{-4}$, is from two separate experimental determinations which gave essentially identical results: one based on a comparison of internal conversion and isomer-shift data [ref. (26)] and a second on comparison of isomer-shift and Knight-shift data [ref. (27)]; both of these determinations avoid ambiguities related to the use of assumptions about electron configurations of tin compounds.

Table 2. Mössbauer Isomer Shifts in the $Au_xSn_y$ system[a])

| | |
|---|---|
| $IS_1 =$ | $+3.5 \pm 0.1$ mm/sec for $^{197}$Au in $AuSn_4$ (absorber relative to an Au absorber), |
| $IS_2 =$ | $-0.57 \pm 0.06$ mm/sec for dilute $^{119}$Sn in Au, (absorber relative to an $AuSn_4$ absorber), |
| $IS_3 =$ | $-0.25 \pm 0.07$ mm/sec for $^{119}$Sn in AuSn (absorber relative to a $\beta$—Sn absorber), |
| $IS_4 =$ | $+3.0 \pm 0.1$ mm/sec for $^{197}$Au in AuSn (absorber relative to an Au absorber). |

a) Ref. (*25*).

with a core electron, whereas the isomer shift depends upon the change only in those conduction electron components having nonzero contact density. In what follows, both sets of data will be utilized to obtain quantitative information about charge transfer in Au—Sn.

**1. Volume Considerations.** Unless an alloy lattice is divided into volumes assigned to the separate constituent atoms, it is impossible to define charge flow between sites. As has been discussed elsewhere (*28*), there is no unique way to make such assignments; however, for the Au—Sn system, a reasonable choice may be based on the volume conservation already mentioned. The atomic volumes in Au and in metallic $\beta$—Sn are 17.0 and 27.0 Å$^3$, respectively. The molecular volume of the ordered compound AuSn is 44.6 Å$^3$, almost exactly the sum of the two pure metal volumes. Similarly, the molecular volume of $AuSn_4$ is 121 Å$^3$, while $4V(\beta\text{—Sn}) + V(\text{Au})$ equals 125 Å$^3$. This close agreement for both compounds suggests that Au and Sn sites in the alloys may well be assigned the pure metal volumes. Isomer and level energy shifts are thus attributable to changes in the charge character within the volumes rather than to changes in the volumes characteristic of the atomic sites. The situation would not be so straightforward if one had only $\alpha$—Sn, the diamond structure, semiconducting allotrope, as the reference material. Because its atomic volume is 34.3 Å$^3$, contraction in the alloy formation would be required, and attribution of the contraction between Au and $\alpha$—Sn sites would be difficult, a priori.

With $\beta$—Sn as the reference, volume effects may be neglected and Eq. (1) may be rewritten

$$\Delta\varepsilon_i = \sum_j^{\text{val}} \Delta\,[n_j I(ij)] - [\sum_j^{\text{val}} \Delta n_j] I_{\text{latt}}\,. \tag{16}$$

Here the $I$'s are core-valence electron interaction terms, of which more will be said later, and $n_j$ is the number of valence electrons $j$. The first term gives the energy shift produced by chemical effects on the parent atom and the second refers to charge changes on the other lattice sites associated with the flow of charge

$$\delta = \sum_j^{\text{val}} \Delta n_j\,. \tag{17}$$

If the chemistry of Sn is considered in terms of $s$ and $p$ band components, Eq. (16) becomes

$$\Delta\varepsilon_i(\mathrm{Sn}) = \Delta n_s[I(i,s) - I(i,p)] + \delta_{\mathrm{Sn}}[I(i,p) - I^{\mathrm{Sn}}_{\mathrm{latt}}] \,, \tag{18a}$$

where Eq. (17) has been used to eliminate $\Delta n_p$. Similarly, transition and noble metals may be described in terms of $d$ and non-$d$ conduction band components, leading to

$$\Delta\varepsilon_i(\mathrm{Au}) = \Delta n_c[I(i,c) - I(i,d)] + \delta_{\mathrm{Au}}[I(i,d) - I^{\mathrm{Au}}_{\mathrm{latt}}] \,. \tag{19a}$$

Using the Mössbauer results to estimate $\Delta n_s$ and $\Delta n_c$, one can attempt to define the net charge flow from the photoemission data.

**2. Isomer Shifts.** Conversion of the data of Table 2 to $\Delta n_s$ for Sn and $\Delta n_c$ for Au must be considered individually for the two atoms because their valence structures differ. Sn is polyvalent but Au is a one-electron metal with filled $d$ bands not far below $\varepsilon_F$.

The Sn $5s$ and $5p$ radial functions, from a nonrelativistic calculation for the free $5s^2 5p^2$ atom, are plotted in Fig. 7. Roughly 8% of the $5s$ charge extends outside the Wigner-Seitz radius, $r_{WS}$, for $\beta$—Sn; the $5s$ orbital, with much of its density in a region in which $Z_{\mathrm{eff}}$ is about equal to the valence, is actually somewhat in the interior of the atom. It is not unlike the $d$ orbitals of transition metals, which, as earlier noted, maintain much of their atomic quality in a metal. Thus it is quite plausible that the valence $s$ character in Sn is much like the free atom $5s$, except for a renormalization within the Wigner-Seitz cell. The much more extended $5p$ component, on the other hand, is not subject to simple renormalization: the $p$ character near the bottom of the band takes on a form more like the dot-dash curve of Fig. 7. It nevertheless appears useful to account for charge terms of a pseudo "$p$" component and a renormalized $s$.

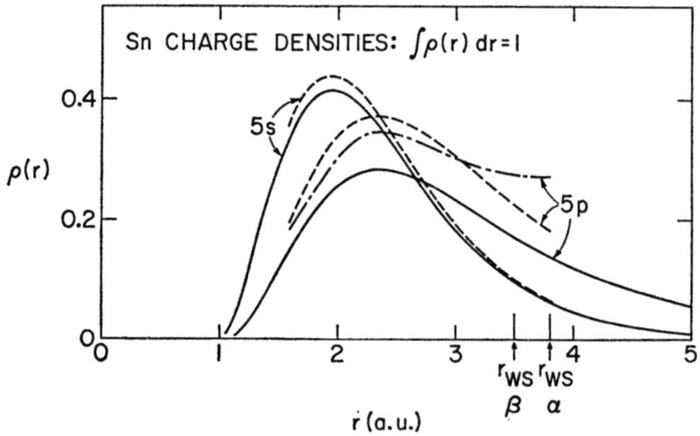


Fig. 7. The valence $5s$ and $5p$ orbital densities of Sn $5s^2 5p^2$. Shown are: free atom densities (solid lines); free atom densities renormalized to the Wigner-Seitz cell of $\alpha$—Sn (dashed curves); and a schematic plot (dot-dash) of the way in which the $5p$ conduction band orbital character deviates from simple renormalization. The Wigner-Seitz radii, $r_{WS}$, of $\alpha$ and $\beta$ Sn are indicated

With the appropriate free atom contact densities, the Sn isomer shifts can now be converted to $\Delta n_s$ values if two factors are accounted for. First, there is the above mentioned 8% modification of the contact density due to renormalization. Second, anticipating the alloy results, which will show that most of the change in $s$ count within the atomic cell is compensated by an opposing $\Delta n_p$, one requires $\Delta n_s$ corresponding to the isomer shift occurring when the atom undergoes a transition $s^x p^y \rightarrow s^{x+\Delta} p^{y-\Delta}$. Employing free atom Fock-Dirac and relativistic Hartree-Fock-Slater calculations, one obtains for the cases of interest

$$\Delta n_s(\beta\text{—Sn} \rightarrow \text{AuSn}) = -\,0.08 \pm 0.02$$

and

$$\Delta n_s(\text{Sn in AuSn}_4 \rightarrow \text{dilute Sn in Au}) = -\,0.17 \pm 0.02\,.$$

In contrast with the 5s in Sn, the 6s electron of a free $d^{10}s$ atom of Au lies outermost, having approximately 40% of its charge outside the $r_{\text{WS}}$ of Au metal. Disortion of this charge density is substantial on forming the metal and discussion of the isomer shift in terms of a modified free-atom-like $s$ orbital might be grotesque. What is more, the natural division for a discussion of Au alloying is between $d$ and non-$d$ effects; therefore one wishes with the isomer-shift data to estimate a count change $\Delta n_c$ in terms of an average conduction-electron charge appropriate to the chemical effects occurring. Lacking anything better, one takes this character to be that of the Fermi-level conduction electrons. The charge-contact density of these electrons may be related to the free atom 6s by comparison of the Knight shift to the free $d^{10}s$ atom magnetic hyperfine constants, *i.e.*, the familiar

$$\xi \equiv \mathscr{K}\,\mu_{\text{B}}/H_{\text{eff}}\chi_{\text{P}}\,, \tag{20}$$

where $\mathscr{K}$, $\mu_{\text{B}}$, $H_{\text{eff}}$, and $\chi_{\text{P}}$ are the Knight shift, the Bohr magneton, the free atom hyperfine field, and the Pauli susceptibility of the metal, respectively. Although electron polarization processes cause magnetic and charge contact densities to differ substantially, it is nonetheless likely that the ratio $\xi$ provides a usable, albeit crude, measure of the ratio appropriate to the charge density. Matters are further complicated by uncertainty in the value of $\chi_{\text{P}}$ for Au and by the fact that one requires a $\xi$ for Fermi-level conduction-band states without contamination by $d$ hybridization. A reasonable estimate based on data from the literature (*29*) is $\xi \simeq 0.6$.

Again, anticipating the alloy results, one wishes to estimate $\Delta n_c$ values for the cases in which the atom remains essentially neutral in charge, *i.e.*, $d^n c^x \rightarrow d^{n-\Delta} c^{x+\Delta}$. This has been done employing free atom calculations and, with a $\xi$ of 0.6, the results are

$$\Delta n_c(\text{Au} \rightarrow \text{AuSn}_4) = +\,0.3 \pm 0.1,$$

and

$$\Delta n_c(\text{Au} \rightarrow \text{AuSn}) = +\,0.26 \pm 0.1,$$

where the uncertainties derive mainly from that in $\xi$.

**3. Core Level Shifts.** Several factors must be considered before the data of Table 1 can be converted into charge shifts. The Au 4*f* and Sn 3*d* spectra display plasmon satellites, which remain constant in relative position and intensity with alloying. Relaxation shifts therefore have been neglected. Rather than to make serious estimates of $\Delta\varepsilon_F$, is is more practical to obtain the charge flow as a function of Fermi level shift for some reasonable range of $\Delta\varepsilon_F$. The work function of Au is ~1 eV greater than of β—Sn: unless the surface barrier term is behaving most unusually, $\Delta\varepsilon_F$ should be $-1 \pm 1/2$ eV on going from Sn to Au.

Let us limit Eqs. (18a) and (19a) to spherical Coulomb interaction terms, $F$'s, between the core electron and the charge external to it. Exchange, if included, would contribute but little in comparison with the several uncertainties. Rewriting (18a) and (19a) as binding energy shifts, we then have

$$-\Delta\mathscr{E}(\mathrm{Sn}) = -\Delta\varepsilon_F + \Delta n_s[F(i,s) - F(i,p)] + \delta_{\mathrm{Sn}}[F(i,p) - F^{\mathrm{Sn}}_{\mathrm{latt}}] \quad (18b)$$

and

$$-\Delta\mathscr{E}(\mathrm{Au}) = -\Delta\varepsilon_F + \Delta n_c[F(i,c) - F(i,d)] + \delta_{\mathrm{Au}}[F(i,d) - F^{\mathrm{Au}}_{\mathrm{latt}}]\,. \quad (19b)$$

It should be noted that the lattice terms in such a set of equations are in general different from each other. If the valence electron orbitals within the cells are frozen, the $F$ functions become the Slater $F°$ integrals, which must be evaluated for valence orbitals normalized to the Wigner-Seitz cell. The use of free atom integrals leads to substantial error, as demonstrated by the values given in Table 3; this is especially the case since differences in Coulomb terms determine $\Delta\mathscr{E}$. Furthermore bare $F°$ integrals do not include valence shell effects which accompany the removal, addition, or change in character of charge in the initial state. This is to say that the terms in the brackets of Eqs. (18b) and (19b) are differentials; they are *not* the Coulomb interactions of some fixed valence orbitals with the core electron in question. These effects can be taken into account by use of effective Coulomb terms, such as

$$\frac{\Delta\mathscr{E}}{\Delta n_d} \equiv F(i,5d) \equiv [\sum F°(i,v)]_{d^n c^x} - [\sum F°(i,v)]_{d^{n-1}c^x} \quad (21)$$

Table 3. Core-valence electron coulomb interaction terms, evaluated from relativistic Hartree-Fock-Slater wave functions

| | $F°$ Bare Integrals | | Screened |
|---|---|---|---|
| | Free Atom (eV) | Renormalized Atom (eV) | Coulomb $F$ Terms (eV) |
| Au $5d^{10}6s$ (4*f*, 5*d*) | 21.2 | 21.7 | 16.1 |
| (4*f*, 6*s*) | 11.5 | 14.3 | 12.8 |
| Sn $5s^25p^2$ (3*d*, 5*s*) | | 15.0 | 13.5 |
| (3*d*, 5*p*) | | 13.1 | 11.9 |

for the removal of a $d$ electron from a Au site. The sums span all valence ($v$) electrons, conduction as well as $d$; the sum in the first bracket involves wave functions for the $d^n c^x$ configuration, the second for the configuration with one fewer $d$ electrons. To calculate the $F^\circ$ of Eq. (21) one may employ renormalized free atom valence orbitals, as has been done here, or orbitals normalized to the W—S cell, which are more appropriate to the solid. A sampling of various pairs of configurations in which the pair members differ by one valence electron results in the terms appearing in the last column of Table 3. Most affected by this effect is the $F(4f, 5d)$ for Au, and consequently the $\Delta n_c$ term of Eq. (19b) is reduced to ~40% of the bare Slater integral value. In any quantitative analysis of core-level shifts one must take into account these initial state differential effects. They are particularly important when one valence shell is well inside another, as is the case for transition and noble metals.

For an ordered intermetallic compound, the $F_{\text{latt}}$ terms of Eqs. (18b) and (19b) may be estimated with a unit charge Madelung type sum. For a dilute impurity, on the other hand, the displaced charge resides in the immediate vicinity and one can make a maximum estimate of $F_{\text{latt}}$ by assuming that the charge is on the surface of the impurity atomic sphere. This leads to $F_{\text{latt}}$ values of 7 eV for Sn as an impurity in Au and 9 eV for Au as an impurity in Sn, the Sn value being smaller because its atomic radius is larger.

With the data accumulated above, Eq. (19b), for the shift from pure Au to Au in $AuSn_4$, becomes

$$-\Delta\mathscr{E}(\text{Au}) = -1.1\ \text{eV} = -\Delta\varepsilon_{\text{F}}(\text{Au} \rightarrow \text{AuSn}_4) + (+0.3 \pm 0.1)[12.8 - 16.1] + \delta_{\text{Au}}[16.1 - 9]$$

or

$$\delta_{\text{Au}}(\text{Au} \rightarrow \text{AuSn}_4) = -0.01 + \frac{\Delta\varepsilon_{\text{F}}(\text{Au} \rightarrow \text{AuSn}_4)}{7} \approx \frac{\Delta\varepsilon_{\text{F}}(\text{Au} \rightarrow \text{AuSn}_4)}{7}.$$

It is obvious that the solution for $\delta_{\text{Au}}$ involves a severe differencing of terms. With $\Delta\varepsilon_{\text{F}}(\text{Au} \rightarrow \text{AuSn}_4)$ taken as $\sim +1$ eV, or the same as the Au—Sn work function difference, the Au charge flow becomes

$$\delta_{\text{Au}}(\text{Au} \rightarrow \text{AuSn}_4) \sim +0.15\ ;$$

similarly, for Sn

$$\delta_{\text{Sn}}(\text{AuSn}_4 \rightarrow \text{dilute Sn in Au}) \sim \frac{\Delta\varepsilon_{\text{F}}(\text{AuSn}_4 \rightarrow \text{Au})}{5} \sim -0.2\,.$$

It happens, in these two calculations, that all terms except those in $\delta$ and $\Delta\varepsilon_{\text{F}}$ have largely cancelled. Cancellation of this kind does not occur with the 50—50 compound; here one has both a simplication — the charge flows at the two sites are necessarily equal in magnitude and opposite in sign — and a complication — greater uncertainty in the $\Delta\varepsilon_{\text{F}}$ values. The result,

$$\delta_{Au}(Au \rightarrow AuSn) = -\delta_{Sn}(Sn \rightarrow AuSn) \sim +0.04 \pm 0.09,$$

carries an error arising mostly from the large uncertainty in $\Delta\varepsilon_F$, which was evaluated on the basis of work function behavior. The Sn $\delta$ values are of the same order and sign as the $\Delta n_s$ inferred from the isomer shifts; although the Au $\delta$'s, too, correspond in sign to the $\Delta n_c$ values, they are substantially smaller than the $\Delta n_c$. It would appear that while the net charge flow is onto Au sites, consistent with its electronegativity, the decrease in $d$ character, which accompanies the increased non-$d$ count, produces a core level shift in the direction usually associated with a loss of charge. This $s$—$d$ compensation has been shown to occur for Au in the Au—Ag and Au—Al systems also (*23, 30*). Depletion of $d$ electrons is not altogether surprising: the $d$ count is less than 10 even in pure Au due to hybridization of the filled $d$ bands with conduction electron states above $\varepsilon_F$; apparently, upon alloying the hybridization increases, presumably involving non-Au sites. Promiscuity of the $d$ electrons may be essential to the large electronegativity of Au.

### B. The Ti—C System

In contrast to the simple additive volume behavior of Au—Sn, there are other intermetallic systems in which the molecular volumes do not correspond to the sum of the atomic volumes characteristic of the constituent elements. Dilute Sb in Ni (*31*) is a case in point: the Sb sites exhibit substantial core level shifts, apparently a concomitant of substantial charge flow. Only small shifts are observed at Ni sites, in part because the solution alloys are quite rich in Ni but also because of the valence $d$ level pinning discussed early in this Section IV. Non-additivity of volume in such systems may indeed be a manifestation of charging.

There are also important classes of metallic compounds in which one component is not metallic in its elemental forms. The transition metal borides and carbides are such compounds, and TiC, which has a simple structure, has been selected for discussion in some detail. Indeed AuSn as well would be such a case if $\alpha$—Sn were the only form of that element. In TiC X-ray absorption and emission and photoelectron results have been obtained for the core levels of both Ti and C. *Ramqvist* and co-workers have analyzed these data by comparing the shifts of different core levels in the same atom (*22*). This approach avoids difficulties caused by uncertainties both in the $\Delta\varepsilon_F$ terms and in $F_{latt}$, but it requires dealing with the small differences between already small individual level shifts. The relevant atomic and molecular volumes are tabulated in Table 4. If diamond, the carbon analog of $\alpha$—Sn, is used as the volume characteristic of C, one has

$$V(\text{hcp Ti}) + V(\text{C-diamond}) = 1.15\ V(\text{TiC})\ ;$$

matters are even worse if graphite is used as the reference, namely

$$V(\text{hcp Ti}) + V(\text{C-graphite}) = 1.30\ V(\text{TiC})\ .$$

Table 4. Atomic and molecular volumes relevant to TiC

| | Volume Å$^3$ |
|---|---|
| $V$ (C-diamond) | 5.67 |
| $V$ (C-graphite) | 8.79 |
| $V$ (bcc Ti) | $18.07_5$ |
| $V$ (hcp Ti) | 17.65 |
| $V$ (TiC) | $20.26_5$ |
| $V$ (hcp Ti) + $V$ (diamond) = | 23.32 |
| $V$ (hcp Ti) + $V$ (graphite) = | 26.44 |

This situation arises, in part, because a dense metallic carbon analog of white $\beta$—Sn does not exist, though the attribution to the Ti sites in TiC of the hcp Ti volume leaves a suspiciously small volume of 2.6 Å$^3$ to the C site. In other words, TiC appears to be compressed to a smaller volume than that allowed by reasonable "metallic" atomic volume assignments for both Ti and C. In what follows it is assumed that

$$V(\text{Ti in TiC}) = V(\text{hcp Ti})/1.15$$

and

$$V(\text{C in TiC}) = V(\text{C-diamond})/1.15\,.$$

The data, summarized in Table 5, show that substantial shifts accompany carbide formation. Although the decrease in binding at C sites, coupled with the increase at Ti, would seem to indicate that charge flows onto C sites, the actual direction of charge flow is probably opposite. As expected, the shift of the Ti 1*s* level with respect to the 2*p* or 3*p* is smaller than the shift of the 2*p* with respect to the Fermi level. Moreover, the 1*s*—2*p*, 3*p* separations are smaller in TiC than in pure Ti. With free atom estimates, *Ramqvist* and co-workers found that the differential Coulomb binding energy shifts associated with change in valence electron count (Table 6) are opposite in sign for $[\mathscr{E}_{1s} - \mathscr{E}_{2p}]$ and $[\mathscr{E}_{1s} - \mathscr{E}_{3p}]$. They concluded that direct chemical effects involving the 3*p* shell perturb

Table 5. Ti and C measured binding energy shifts (in eV)[a]

| | |
|---|---|
| $\Delta\mathscr{E}_{1s}$ (C; graphite → TiC) | −3.3 |
| $\Delta\mathscr{E}_{2p}$ (Ti; Ti → TiC) | +1.3 |
| $\Delta\mathscr{E}_{1s} - \Delta\mathscr{E}_{2p}$ (Ti; Ti → TiC) | −0.33 |
| $\Delta\mathscr{E}_{1s} - \Delta\mathscr{E}_{3p}$ (Ti; Ti → TiC) | −0.44 ± 0.1 |

[a]) Ref. (*22*).

Table 6. Calculated differential Coulomb binding energy shifts due to change of valence $s$ and $d$ count in Ti[a])

| | $\Delta[\mathscr{E}_{1s}-\mathscr{E}_{2p}]$ | | $\Delta[\mathscr{E}_{1s}-\mathscr{E}_{3p}]$ | |
|---|---|---|---|---|
| | Free atom (eV) | Renormalized to volume of Ti in TiC (eV) | Free atom (eV) | Renormalized to volume of Ti in TiC (eV) |
| $d$ gain $d^2s \to d^3s$ | +0.54 | +0.39 | −0.39 | −2.05 |
| $d \to s$ $d^3s \to d^2s^2$ | −0.62 | −0.95 | +0.34 | +0.34 |

[a]) Relativistic Hartree-Fock-Slater calculations were used to derive the results in this Table and in Tables 7 and 8. For free atoms, these are simply obtained from differences between core orbital one electron energies computed for various valence configurations. For atoms confined to Wigner-Seitz volumes, the core-valence Coulomb terms used were based on free atom valence electron orbitals renormalized to the appropriate atomic volumes. Effects on core-valence exchange were neglected.

$[\mathscr{E}_{1s}-\mathscr{E}_{3p}]$. It is not necessary, however, to invoke any such specific effect. In differential form

$$\Delta[\mathscr{E}_{1s}-\mathscr{E}_{np}] = \Delta V\left\{\frac{\Delta[\mathscr{E}_{1s}-\mathscr{E}_{np}]}{\Delta \mathrm{Vol}}\right\} + \Delta n_c\left\{\frac{\Delta[\mathscr{E}_{1s}-\mathscr{E}_{np}]}{\Delta n_{d\to s}}\right\} + \delta\left\{\frac{\Delta[\mathscr{E}_{1s}-\mathscr{E}_{np}]}{\Delta n_d}\right\}, \tag{22}$$

where, like in the case of Au,

$$\delta = \Delta n_c + \Delta n_d$$

and the three terms arise from partial differential effects on the binding energy difference of the volume change, the $d$ to $s$ valence count change, and the net charge flow. Because Eq. (22) is written not for binding energy shifts directly, but for shifts of a difference between core level binding energies in a given atom, neither the $\Delta\varepsilon_F$ nor the lattice terms in expressions such as Eqs. (18b) and (19b) appear. Moreover, since to first approximation relaxation effects occurring outside the core are independent of the particular orbital in which the core vacancy is made, no final state relaxation terms appear in Eq. (22). Two Eqs. (22) written for the 1$s$—2$p$ shift and for the 1$s$—3$p$ shift may be solved simultaneously; if one substitutes values from Table 7 for the first terms and from Table 6 for the second term differentials (configuration $d^3s$ was used), one finds that

$$\delta(\mathrm{Ti} \to \mathrm{TiC}) = +0.06 \pm 0.05$$

and

$$\Delta n_c(\mathrm{Ti} \to \mathrm{TiC}) = +0.3 \sim -\Delta n_d\,.$$

These results show that on formation of TiC there is perhaps a small flow of electron charge onto Ti sites and that there is significant conversion in the valence shell of $d$ to non-$d$ character, as was observed to occur at Au sites in the Au—Sn system.

Table 7. Coulomb shifts of Ti and C binding energies due to change of atomic volume[a)]

| | | |
|---|---|---|
| C | $2s^22p^2$ | $2s2p^3$ |
| $\frac{\Delta\mathscr{E}\ 1s}{\Delta\ \mathrm{Vol}}\ \Delta V\,(\text{graphite}\rightarrow\text{TiC})$[b)] | $-6.9$ eV | $-7.2$ eV |
| Ti | $3d^34s$ | |
| $\frac{\Delta\mathscr{E}\ 2p}{\Delta\ \mathrm{Vol}}\ \Delta V\,(\text{hcp Ti}\rightarrow\text{TiC})$ | $-1.3$ eV | |
| $\frac{\Delta[\mathscr{E}_{1s}-\mathscr{E}_{2p}]}{\Delta\ \mathrm{Vol}}\ \Delta V\,(\text{hcp Ti}\rightarrow\text{TiC})$ | $-0.07$ | |
| $\frac{\Delta[\mathscr{E}_{1s}-\mathscr{E}_{3p}]}{\Delta\ \mathrm{Vol}}\ \Delta V\,(\text{hcp Ti}\rightarrow\text{TiC})$ | $-0.42$ | |

a) See footnote Table 6.
b) Terms are defined in analogy with the example

$$\frac{\Delta\mathscr{E}_{1s}}{\Delta\ \mathrm{Vol}}\ \Delta V\,(\text{graphite}\rightarrow\text{TiC} \equiv -\left\{\left[\sum_{\text{val}} F^\circ(1s,\ \text{val})\right]_{\text{TiC}} - \left[\sum_{\text{val}} F^\circ(1s,\ \text{val})\right]_{\text{graphite}}\right\}.$$

Independent estimates of $\delta$ and $\Delta n_c$ may be had from the observed C 1*s* and Ti 2*p* photoemission line shifts themselves; the equations analogous to (18b) and (19b) for these shifts are

$$\begin{aligned}\Delta\mathscr{E}_{1s}(\text{C};\ \text{graphite}\rightarrow\text{TiC}) &= -\ 3.3\\ &= \Delta\varepsilon_{\text{F}}(\text{C}) + \frac{\Delta\mathscr{E}_{1s}}{\Delta\ \text{Vol}}\ \Delta V\,(\text{graphite}\rightarrow\text{TiC})\\ &\quad - \Delta n_s[F(1s,\ 2s) - F(1s,\ 2p)]\\ &\quad + \delta\,[F(1s,\ 2p) - F^{\text{C}}_{\text{latt}}]\\ &= \Delta\varepsilon_{\text{F}}(\text{C}) + [-\ 7] - \Delta n_s[-\ 0.7] + \delta\,[12.8]\end{aligned} \tag{23}$$

and

$$\begin{aligned}\Delta\mathscr{E}_{2p}(\text{Ti};\ \text{Ti}\rightarrow\text{TiC}) &= 1.3\\ &= \Delta\varepsilon_{\text{F}}(\text{Ti}) + \frac{\Delta\mathscr{E}_{2p}}{\Delta\ \text{Vol}}\ \Delta V\,(\text{hcp Ti}\rightarrow\text{TiC})\\ &\quad - \Delta n_c[F(2p,\ 4s) - F(2p,\ 3d)]\\ &\quad - \delta[F(2p,\ 3d) - F^{\text{Ti}}_{\text{latt}}]\\ &= \Delta\varepsilon_{\text{F}}(\text{Ti}) + [-\ 1.3] - \Delta n_c[-\ 5.6] - \delta[8.2]\,,\end{aligned} \tag{24}$$

where

$$(\Delta n_c + \Delta n_d)_{\text{Ti}} \equiv \delta \equiv -\ (\Delta n_s + \Delta n_p)_{\text{C}}$$

and where the quantities in the square brackets[4)] are taken from Tables 7 and 8.

4) The Madelung sum derived $F_{\text{latt}}$ value was employed. Values obtained with the assumption that the displaced charge resides at the surface of the atomic sphere and with the neglect of charge in the remainder of the lattice are also tabulated in Table 8. Their average is in excellent agreement with the Madelung result.

The $\Delta\varepsilon_F$(Ti) value is obtainable from the data in Table 9. Graphite has been used for the Fermi level reference in Eq. (23). Unless $\Delta n_s$ in Eq. (23) is an order of magnitude grater than $\delta$, which is unlikely, its term is small enough to be dropped; and the equations can then be solved to yield

$$\delta(\mathrm{Ti} \rightarrow \mathrm{TiC}) = +0.2$$

and

$$\Delta n_c(\mathrm{Ti} \rightarrow \mathrm{TiC}) = +0.6\,.$$

Table 8. Effective core-valence electron Coulomb interaction energies and $F_{\mathrm{latt}}$ term energies for Ti and C. (Values in eV)[a]

| | | | Free atom | Graphite volume | TiC volume |
|---|---|---|---|---|---|
| C | $F(1s, 2s) - F(1s, 2p)$ | $sp^3 \rightarrow s^2p^2$ | + 0.7 | − 0.4 | − 0.7 |
| | | $sp^2 \rightarrow s^2p$ | − 0.2 | − 0.7 | − 1.0 |
| | $F(1s, 2p)$ | $s^2p \rightarrow s^2p^2$ | 15.0 | 21.4 | 24.4 |
| | | $sp^2 \rightarrow sp^3$ | 14.1 | 21.2 | 24.2 |
| Ti | $F(2p, 4s) - F(2p, 3d)$ | $d^4 \rightarrow d^3s$ | − 2.6 | | − 5.9 |
| | | $d^3s \rightarrow d^2s$ | − 5.4 | | − 5.3 |
| | $F(2p, 3d)$ | $d^2s \rightarrow d^3s$ | 12.5 | | 19.2 |
| | | $ds^2 \rightarrow d^2s^2$ | 15.3 | | 20.2 |
| | $F^{\mathrm{Ti}}_{\mathrm{latt}} = F^{\mathrm{C}}_{\mathrm{latt}}$ | 11.5 | From Madelung sum | | |
| | $F^{\mathrm{Ti}}_{\mathrm{latt}}$ | 9.3 | From assumption that displaced charge | | |
| | $F^{\mathrm{C}}_{\mathrm{latt}}$ | 13.5 | is at surface of atomic sphere | | |

[a]) See footnote Table 6.

Table 9. Work functions for the Ti—C system[a])

| | |
|---|---|
| Ti | 4.1 eV |
| graphite | 4.4 |
| TiC | $3.2 \pm 0.8$ |

[a]) *Rivière, J. C.*: Solid State Surf. Sci. *1*, 179 (1969) and *Fomenko, V. S.*: Handbook of thermionic properties (ed. by *G. V. Samsonov*). New York: Plenum 1966.

This set of values and the set derived by use of Eq. (22) correspond in signs. Although these values are somewhat large relative to the others, they would be decreased if the volume term of Eq. (23) were made smaller in magnitude as it probably would be if intraatomic relaxation about the final state hole were taken into account. It is unlikely, however, that any computational refinements would go so far as to change signs of $\delta$ and $\Delta n_c$ deduced using graphite as the reference material.

Two observations should be made concerning the above sets of values. First, the choice of Ti and C volumes in TiC is somewhat arbitrary. As discussed earlier, it would be plausible to ascribe a larger fraction of the molecular volume to the Ti site, with a resulting increase in the magnitude of $\delta$. Second, relaxation effects have been ignored in Eqs. (23) and (24) at the expense of some error. One expects the valence electrons to be more effective in screening a carbon hole in TiC than in graphite; they are perhaps somewhat less effective in screening a Ti site in TiC than in Ti metal. In other words, relaxation should make a negative contribution to $\Delta\mathscr{E}$ (graphite $\rightarrow$ TiC) and a positive one to $\Delta\mathscr{E}$(Ti $\rightarrow$ TiC). If one thus adjusts the observed shifts to obtain the shifts without relaxation for which Eqs. (23) and (24) are written, these equations lead to further increased values for $\delta$ and $\Delta n_c$.

For both Ti and C in TiC as well as for the case of Au in Au—Sn the direction of the charge flow appears to be opposite to that suggested by simple inspection of the observed shifts in individual core levels. The $d \rightarrow s$ terms were seen to be important in Au and the volume effects critical in TiC. The result for TiC is uncertain because it depends on the large volume terms of Table 7, in the calculation of which intraatomic relaxation effects due to squeezing of the valence electrons have been neglected, of necessity. Diamond rather than graphite may be used as the carbon reference material; the volume correction then is smaller, but the Fermi level term, $\Delta\varepsilon_F$, is unknown and extraatomic relaxation effects, neglected in Eq. (23), may be significant. On balance it would appear that the charge flow is small, $\lesssim 0.1$ e, and quite likely is from C to Ti, a direction opposite to that expected from conventional electronegativity considerations. A contrary result, indicating that substantial charge flow does occur on carbide formation, was arrived at by Ramqvist through neglect of all terms other than the $\delta$ term in Eq. (23).

Measurement of shift in one core level with respect to another in the same atom avoids problems of evaluating terms in $\Delta\varepsilon_F$ and $F_{latt}$ and lessens the importance of relaxation; these gains are made, however, at the cost of more severe numerical differencing among those terms which remain. Either of the two approaches can yield useful semiquantitative information on chemical effects in an alloy sequence. Investigations of this type would be facilitated if there were available data on useful reference quantities for atomic species of interest. For example, it would be valuable to have for C a "metallic" atomic volume, and associated Fermi level energy, and the position of the 1s core level with respect to that $\varepsilon_F$ so that these quantities could be used as references for C in the metal carbides. Perhaps as this subject outgrows its infancy, the criteria for such a reference catalog and in turn the catalog itself will be developed.

## V. Conduction Bands

### A. Pure Metals

As a rule in X-ray photoemission studies, more effort has been expended on the conduction bands of a metal and its alloys than on its core levels. Because the Fermi level of course lies in the bands, band level energies have not so much been the object of investigation as band structure has been; and it is considered that the photoelectron spectrum provides a reasonably faithful reproduction of this structure — the density of occupied states as a function of electron energy, $\varrho(\varepsilon)$. Implicit in this consideration is the assumption that the final state of the photo-excited electron lies high in an essentially structureless part of the continuum. Nevertheless, variation of transition probability across the bands does affect the spectrum; at soft X-ray photon energies, *d*- and *f*-like band components, for example, have larger transition matrix elements than *s* and *p* components. *Eastman* and colleagues (*32*), in studies of Au photoemission at various photon energies in the ultraviolet, have provided insight into these matters. Some of their results appear in Fig. 8. It is generally accepted that the initial and final states of the photoelectron must have the same wave vector $\boldsymbol{k}$. For a photon energy, $h\nu$, the photoelectron intensity at energy $E$ is then

$$I(E) = \int\int_0^{\varepsilon_F} \varrho(\varepsilon, \boldsymbol{k})\, \varrho_c(E, \boldsymbol{k})\, |T(\varepsilon, E, \boldsymbol{k})|^2\, \delta(E-\varepsilon-h\nu)\, d\boldsymbol{k}\, d\varepsilon\,, \tag{25}$$

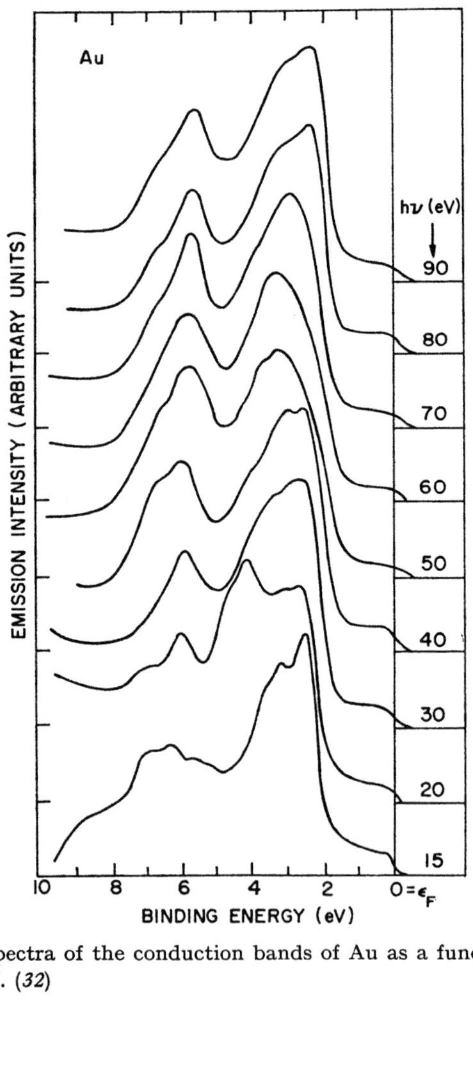


Fig. 8. Photoelectron spectra of the conduction bands of Au as a function of photon energy ($15 \leq h\nu \leq 90$ eV). Ref. (*32*)

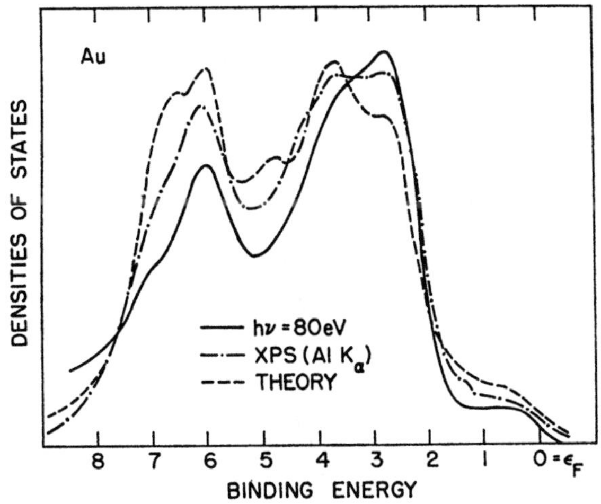


Fig. 9. X-ray and UV photoelectron spectra of the conduction bands of Au and a theoretical density of states curve. For purposes of comparison the theoretical curve includes broadening with a 0.6 eV resolution function. Ref. (*32*)

where $T$ is the transition matrix element, $\varrho$ is the density of states in the occupied bands, and $\varrho_c$ is the density of states in the continuum above the Fermi level. Figure 8 shows that when the photon energy has increased to 50 eV or thereabouts the spectra have very nearly attained a final high energy form, much as if $\varrho_c$ has become "white" and is no longer making any contribution to the structure of $I(E)$. In Fig. 9, an XPS spectrum is compared with the 80 eV UPS curve and with a theoretical density of states, appropriately broadened by combination with a 0.6 eV resolution function. The three curves are in semiquantitative agreement. When the high energy peaks are normalized to one another, the experimental curves lie under theory at the lower energies. This is almost undoubtedly due to the behavior of the transition probability as discussed below. Another comparison is made in Fig. 10 for Cu (*33*). In this case the XPS result and a soft X-ray emission (SXS) spectrum are shown. Except for the low energy tail and the relatively poor resolution, both characteristic of SXS spectra, the two sets of measurements are in moderately good agreement. When structure in the form of a peak, bump, or shoulder is found in an XPS, UPS, or SXS spectrum of a transition or noble metal, it will generally be found in all three spectra, perhaps with different strengths. This is somewhat surprising, in view of the different selections of initial and final states in the three experiments, particularly with SXS, which involves an initial core hole state.

Both *Baer* and co-workers (*34*) and *Fadley* and *Shirley* (*35*) have studied a number of the transition metals. The Uppsala results are plotted in Fig. 11. Doublets are clearly observed for Ir, Pt, and Au; the other spectra consist of single peaks, each with its characteristic width and hints of structure. In XPS, as a rule, intensity at the bottom of *d* bands and the indications of structure there are suppressed, apparently because of transition probability decrease. Hybridization of *d* with non-*d* character, which is generally strong toward the bottom of the *d* bands and weak at the top, tends to lessen transition intensity. It is suggestive

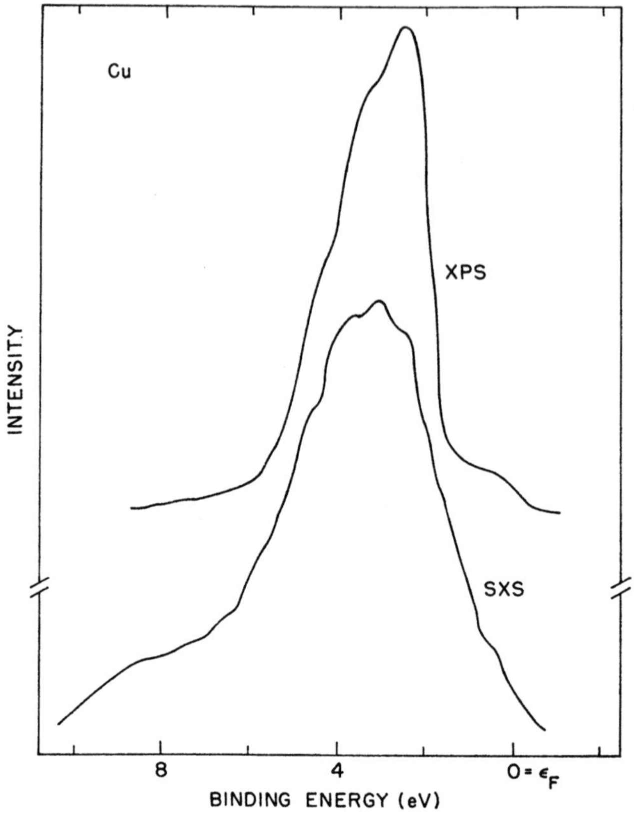


Fig. 10. Comparison of X-ray emission (SXS) and X-ray photoelectron (XPS) spectra for the conduction bands of copper. Ref. (*33*)

that the three metals having the strongest *d* electron spin-orbit effects are those which show doublet-like structures most clearly. Splitting in these structures, caused by band effects as well as spin-orbit interaction, does not match the free atom $\boldsymbol{l} \cdot \boldsymbol{s}$ separation. In Au, for example, it is 3 eV, almost twice the atomic $d_{3/2}$—$d_{5/2}$ difference. Relative intensities of soft X-ray emission lines indicate, nevertheless, that the states associated with the peaks are each predominantly of an individual $j$ value; this fact seems to be connected with the lack of suppression of intensity of the lower lying peak.

As vacuum systems and spectrometers are improved, data such as those of Fig. 11 are being obtained with increasingly high resolution and metals of increasing reactivity are being studied. Some recent results of *Baer* and *Busch* for Gd, Dy, and Er are shown in Fig. 12 (*36*). Only the conduction band regions are plotted; broadened theoretical $\varrho(\varepsilon)$ curves for comparison are on the right. Rather good agreement between theory and experiment is seen in the trends along this sequence of metals, which like Sc and Y are transition elements with partially filled bands in approximately a $5d^2 6s$ configuration. Unlike the heavy transition metals of Fig. 11, one is scanning only a small portion of the *d* bands here.

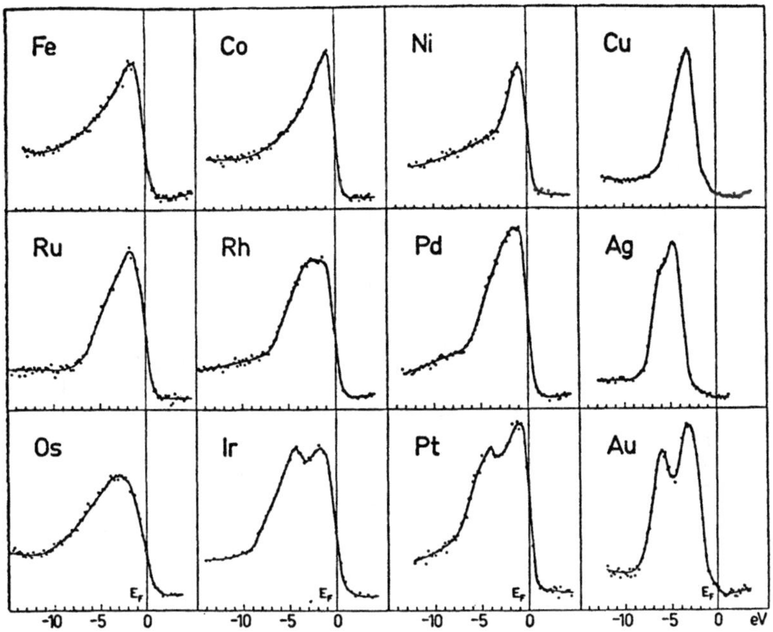


Fig. 11. X-ray photoemission spectra of the conduction bands in the heavier members of the 3*d*, 4*d*, and 5*d* transition metals. Ref. (*34*)

## B. Alloys

According to the so-called rigid band description of alloy conduction bands, the different atomic species present donate their valence electrons to a common band structure. The "virtual crystal" model, for example, incorporates this description. On the other hand, we have noted that the center of gravity of the *d* bands in a transition or noble metal is determined by an atom-like one-electron energy $\varepsilon_d$. This raises the interesting question as to whether, in an alloy consisting of two transition metals, there are in first approximation separate *d* bands associated with the individual atomic species, each in the vincinity of its $\varepsilon_d$. In second approximation, these might to some extent be hybridized with one another. The "coherent potential" approximation, for example, incorporates this view. Experiments with UPS and XPS have been carried out to discriminate between such models. The clearest cases for consideration are those in which the energies, $\varepsilon_d$, for the constituents are well separated; this is the situation when, for example, one constituent is a noble metal and the other a transition metal. For the Cu—Ni (*37*) and Ag—Pd (*38*) systems, the XPS results of *Hüfner et al.* show, in Figs. 13 and 14 respectively, that there indeed are individual *d* bands corresponding closely to those of the pure constituents. Shifts of the $\varepsilon_d$ values relative to $\varepsilon_F$ with change of composition are only slight; but as a constituent becomes more dilute and the distance between atoms of the species increases, its *d* band narrows, in accord with expectation. In the limit of infinite dilution, the bandwith does not go to zero because conduction-*d* electron scattering contributes to the *d* level width

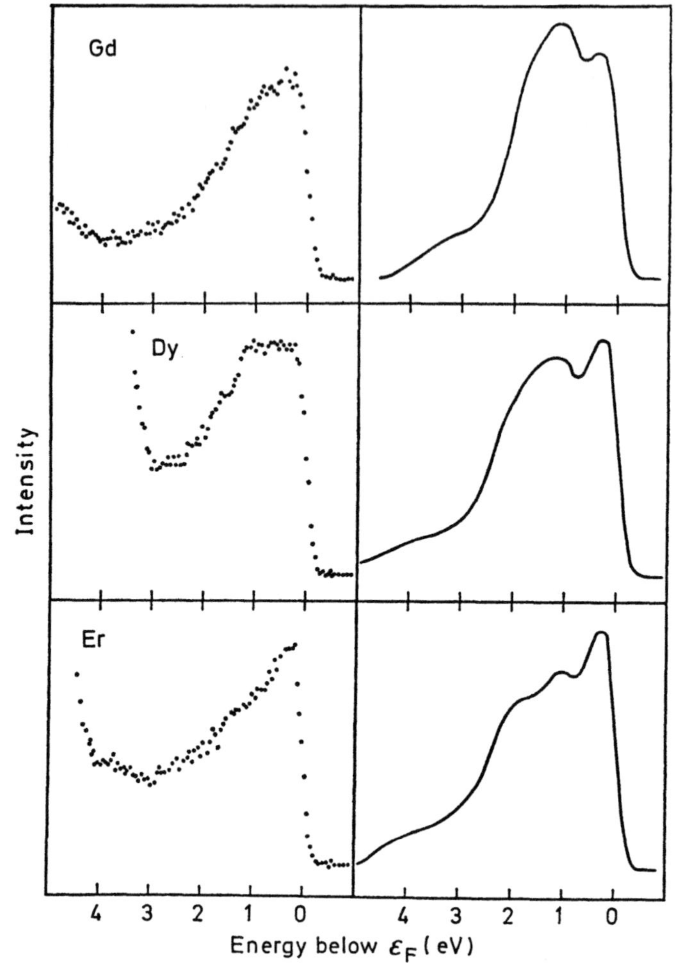


Fig. 12. High resolution photoemission spectra of the conduction electrons of Gd, Dy, and Er in comparison with theoretical densities of states (solid lines) which have been broadened so as to reproduce the finite experimental slopes at $\varepsilon_F$. Ref. (*36*)

associated with even an isolated site. To test this point *Hüfner et al.* used the Ag—Pd results, better suited for the purpose than the Cu—Ni data because the two $\varepsilon_d$ are more widely separated and because the measurements were made at higher resolution. By extrapolation to infinite dilution, they obtained a *d* level width of ~1/2 eV for either isolated Pd or Ag, a result, as they observed, in accord with predictions based on the coherent potential approximation. Even before the UPS (*39*) and XPS results were available it had been widely held that bands of the constituents tend to maintain identity in alloying; data such as those shown in Figs. 13 and 14 provide visual proof of this.

One can also ask what is the *d* band behavior when only one of the constituents of a binary alloy has such bands. The spectra plotted in Fig. 15 demonstrate this

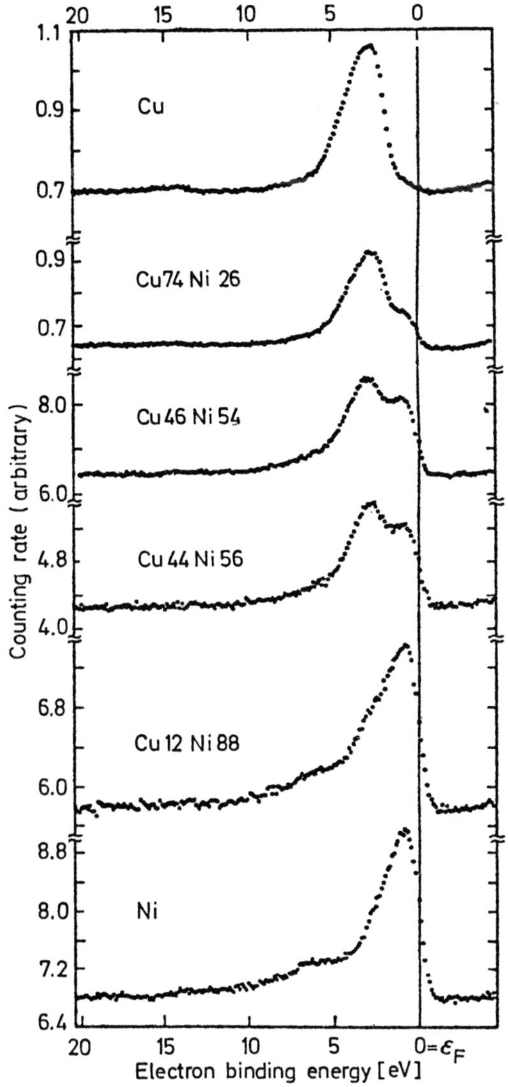


Fig. 13. X-ray photoelectron spectra of the valence bands in the Cu—Ni system. Ref. (*37*)

behavior; they are for Au—Sn, the system whose core level shifts were considered in Section IV above. It is evident that there is a substantial shift in $\varepsilon_d$, relative to $\varepsilon_F$, upon alloying, a shift in the same direction as, but greater than, that shown by the core levels.

This could not readily be understood if the *d* bands were merely sampling the changing potential at the Au site. The point, of course, is that the Au *d* bands are *actively* involved in the alloying, as shown by the decrease in *d* count deduced from the core level and Mössbauer measurements. With dilution of the Au, the *d* bands become increasingly narrow and in $AuSn_4$ the peak separation has decreased to

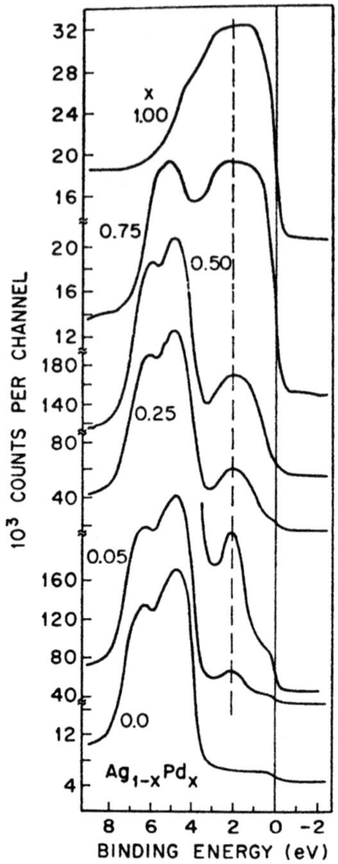


Fig. 14. X-ray photoelectron spectra of the valence bands in the Ag—Pd system. Ref. (*38*)

approximately the free atom $d_{3/2}$—$d_{5/2}$ splitting. Simple increase of distance between Au atoms should eventually lead to a *d* band splitting the same as the spin-orbit splitting of isolated free atoms. In the dilute solid alloys, however, any hybridization of the *d* band electrons with the Sn orbitals, by reducing the time spent by the electrons at the Au site, reduces the spin-orbit interaction and thus could lead to observed splittings smaller than that for the free atom. In $AuSn_4$, the most dilute alloy measured in these experiments, the doublet splitting corresponds to that of the free atom and hence provides no evidence for *d* hybridization with Sn. There is, nevertheless, an indication of hybridization in the $AuSn_4$ spectrum. The upper edge of the *d* band spectrum for pure Au and the other alloys drops off quite sharply; the drop-off is much less sharp in $AuSn_4$. This is probably an effect of *d* hybridization with either Sn or with Au site conduction electron character at the upper edge of the *d* bands. Such hybridization would show itself, in addition, by a decrease in intensity of the main *d* peaks.

Au is notable for the strikingly unusual colors of its intermetallic compounds. $AuAl_2$, for example, is mauve. Such a color is a consequence of pronounced

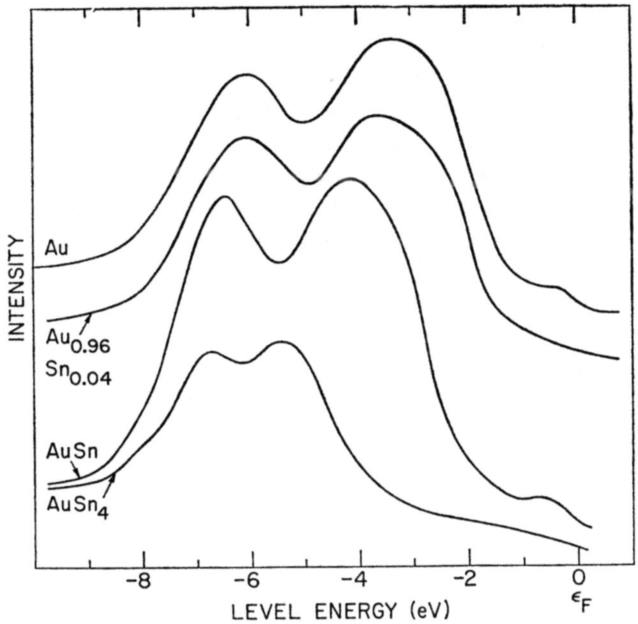


Fig. 15. X-ray photoelectron spectra of the valence bands of pure Au, dilute Sn in Au, and two Au—Sn compounds. Ref. (*23*)

structure in the states at the upper edge of the Au *d* bands or in the valence bands above $\varepsilon_F$ or both. Results such as those for $AuSn_4$ indicate that, at the least, hybridization occurs in that part of the *d* band which is linked to color. However, the role of hybridization in color production as well as in the extreme electronegativity of Au needs further investigation; in this XPS studies should be quite useful[5]).

Although many XPS investigations have been made of valence bands of systems in which the constituents are elements other than the transition and noble metals, there has been relatively little such work on metallic alloys of such elements. This situation will undoubtedly change.

## VI. Substructure in Line Spectra

### A. Core Level Structure

In first approximation, one might expect to observe a single electron line from the photoexcitation of a core *s* level and a spin-orbit doublet from levels of higher *l*. Line widths, in this approximation, would be a function of the core vacancy lifetimes set by the various filling mechanisms. As mentioned earlier, the core shell

---

[5]) Zintl compounds, many of which are colored, should be of interest for these studies. Unfortunately they are extremely reactive.

of highest $n$ and $l$ tends to have the narrowest lines and it is therefore generally chosen for measurement of chemical shifts.

It is clear, however, from the discussion involving Eqs. (7)—(9) and from the sudden approximation sum rule that the spectrum associated with the photoionization of a core electron should not, in fact, necessarily consist only of a single line; some data observed for RbCl and RbF (*40*) are shown in Fig. 16. The narrow peaks are the Rb $4s^24p^6(^1S) \rightarrow 4s^14p^6(^2S)$ excitation and the broad peaks, approximately equal in intensity, arise from "multiple electron excitation", that is, the production of final states such as $4s^24p^4ns(^2S)$, where $n \geq 5$. Even though the photoemission event is just a one-electron dipole process, "multiple excitation" can occur because the wavefunction $\Phi$i of the "instantaneous" intermediate state of the ($N$—1)-electron ion [Eq. (7)] has overlap with wavefunctions of such multiply excited states; that is, $\Phi_i$ has components which are eigenfunctions, $\Psi_n(N—1)$, of multiply

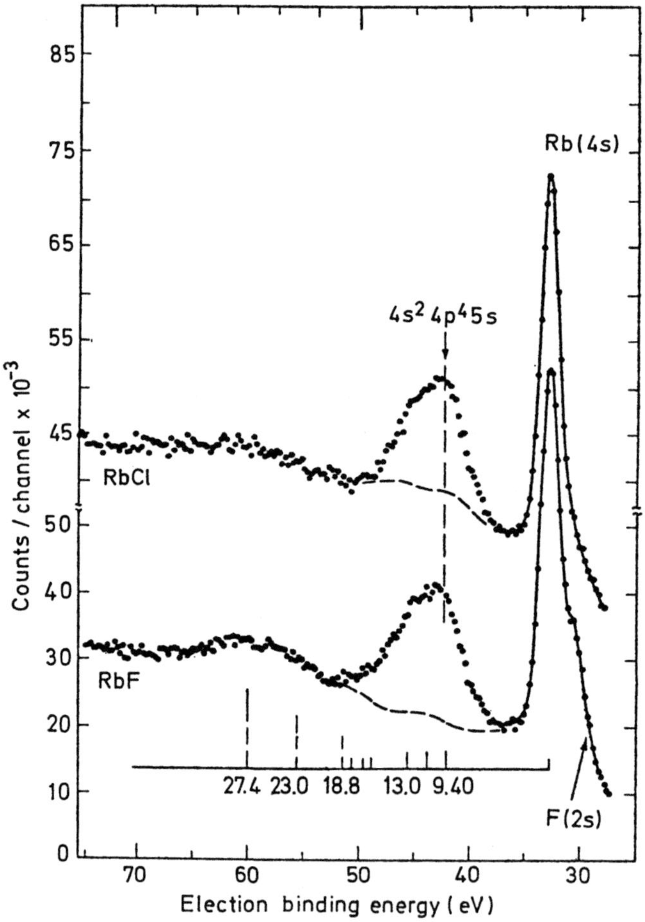


Fig. 16. X-ray photoelectron spectra of Rb 4$s$ in RbCl and RbF; shakeup effects are responsible for the satellites. The position of the $4s^24p^45s\,(^2S)$ level, 9.56 eV more tightly bound than the $4s^14p^6\,(^2S)$ state represented by the sharp peak in each case, is indicated (see text). Ref. (*40*)

excited states. Non-zero overlap[6]) may occur when there is configuration mixing (correlation effects) in either the initial state $\Phi_0(N)$ or the final state $\Psi_n(N-1)$. Consider the latter for the case at hand: in configuration mixing notation, the $4s^14p^6(^2S)$ determinant associated with the final state is mixed with wavefunctions of like symmetry, such as $4s^24p^45s(^2S)$, and thus there are transitions to the several final states sharing $4s^14p^6(^2S)$ character. The relative position of the $4s^24p^45s(^2S)$ level, as estimated from atomic spectra, is indicated on the figure. When the final state of the second excited electron is bound, a "shakeup" process is said to be involved; when it is unbound, "shakeoff" has occurred. These processes are not to be confused with the energy loss, due to scattering, which may be sustained by a photoelectron as it escapes the sample. Shakeoff and shakeup have been observed and analyzed for the inert gases in work done at Oak Ridge (*41*) and at Uppsala (*2*). In contrast with the results obtained from the Rb halides, a set of narrow satellites was observed to arise from the excitation of the second electron to energetically well defined unoccupied atomic levels. In an insulating solid, such as RbCl or CuO, whose spectrum (*42*) appears in Fig. 17, the satellites tend to appear in a broad peak which sometimes has in it indications of structure. The valence bands lying above $\varepsilon_F$ in such a material allow, for the second electron, a distribution of excitations, which leads to a broad peak. In metals such as Cu (Fig. 17) and Ni these broad satellite peaks are not observed; it may be that second electron excitations to regions above $\varepsilon_F$ have little overlap with $\Phi_i$ or that, because the bands above $\varepsilon_F$ are broad, the satellite intensity is also broad and thus low and hidden in the background. Other weaker satellite peaks associated with plasmon excitation, collective modes of the conduction electrons, are generally visible in the spectra of metals (see note added in proof).

Multiple line core spectra are produced also if the atom has an open valence shell, provided that the crystalline environment has not wiped out the $J, L, S, M$ quantization of that shell; the core vacancy is variously coupled to the open shell to yield a set of final states. For example, if the open shell has the one-electron orbital quantum numbers $n$ and $l$ and total spin $S$, a core $s$ vacancy will be observed in two final states having spins $(S+1/2)$ and $(S-1/2)$, with the latter spin state lying higher in energy. According to Condon-Slater-Racah theory, the energy separation is

$$\Delta E = \frac{(2S+1)}{(2l+1)} G^l(s, nl), \tag{26}$$

where $G^l$ is a Slater exchange integral

$$G^l(i, j) \equiv \iint P_i(r_1)P_j(r_1) \frac{r_<^l}{r_>^{l+1}} P_i(r_2)P_j(r_2)r_1^2dr_1r_2^2dr_2. \tag{27}$$

[6]) Note that the sum rule, Eq. (9c), holds if there are correlation effects associated with the final state hole but it cannot, by definition, hold if correlation is important in the initial state.

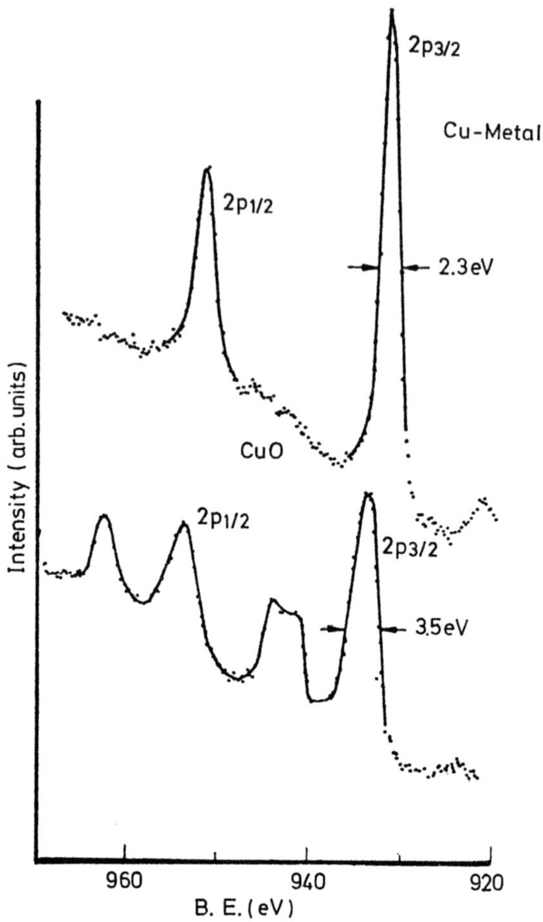


Fig. 17. The $2p$ X-ray photoelectron spectra of Cu in CuO and Cu metal. Shakeup satellites appear in the oxide spectrum but not in that of the metal. Ref. (*42*)

Here the $P$'s are radial wavefunctions associated with orbitals $i$ and $j$; and

$$r_< = r_1 \text{ and } r_> = r_2 \text{ if } r_2 > r_1$$

but

$$r_< = r_2 \text{ and } r_> = r_1 \text{ if } r_2 < r_1 .$$

The relative intensities of the two lines are expected to be proportional to the final state degeneracies, *i.e.*,

$$\frac{I(S+1/2)}{I(S-1/2)} = \frac{2(S+1/2)+1}{2(S-1/2)+1}, \tag{28}$$

if the transition probabilities are equal.

When a core hole of high $l$ is involved, the spectrum can become quite complicated, as is the case for the $4d$ excitation of Sm metal (Fig. 18). Although such

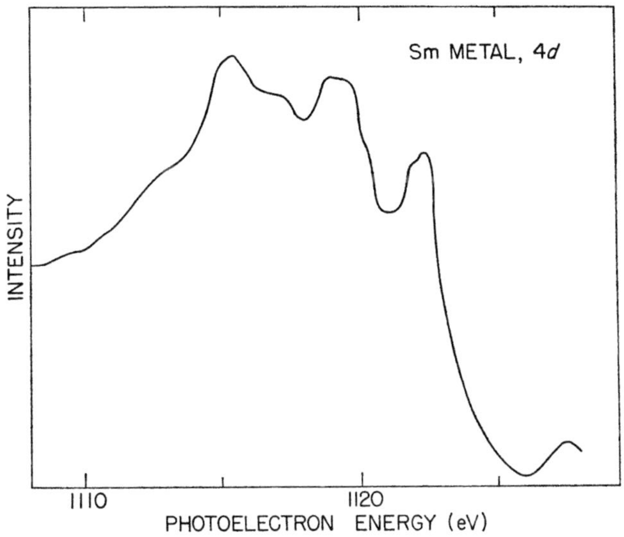


Fig. 18. The 4*d* X-ray photoelectron spectrum of Sm metal (J. Cuthill, unpublished)

a spectrum has yet to be analyzed, it may be noted that the rightmost peak, corresponding to the final state of lowest energy, is in first approximation the Hund's Rule ground state of the $4f^54d^9$ configuration; and chemical shifts of the 4*d* level can be analyzed in terms of the motion of this ground level rather than from estimates of the motion of the centroid of the full 4*d* spectrum.

*Fadley* and co-workers (43) studied the *s* shell excitations of $Mn^{2+}3d^5(^6S)$ in $MnF_2$ and discovered that the splitting of the 2*s* peak conforms to the Eqs. (26)—(27) result but that the 3*s* splitting was much smaller than the calculated value. Similarly, for the rare earth ions in salts, the 5*s* shell splittings correspond to Eqs. (26)—(27), but the 4*s* splittings correspond to Eq. (26) only if the Slater $G^3(4s, 4f)$ parameters are substantially smaller than the values of the $G^3$ integrals given by Hartree-Fock theory (*44, 18*). Now, Hartree-Fock theory is known to overestimate the stability of a magnetic level, such as a state of maximum spin multiplicity, relative to that of a magnetically weaker level. The exchange interaction between electrons of like spin, included in the Hartree-Fock method, substitutes for some of the interelectronic "correlation" among them. When the electrons are of antiparallel spin, interelectronic exchange is absent and the correlation effects omitted from Hartree-Fock theory are more important energetically. As a result, splitting ratios in the optical multiplet spectra of valence shells of atoms are well described by expressions such as Eq. (26), but to obtain the magnitudes of the splittings one needs the screened Slater *parameters*, which are indeed smaller than the Slater *integrals*. Photoionization of the 2*s* electrons of $Mn^{2+}$ and the 5*s* of the rare earth ions are both excitations of core levels for which the Slater parameter values inferred from experiment agree with the Slater integrals. This indicates that correlation effects are, in these instances, unimportant presumably because of the absence of near lying levels available for configuration

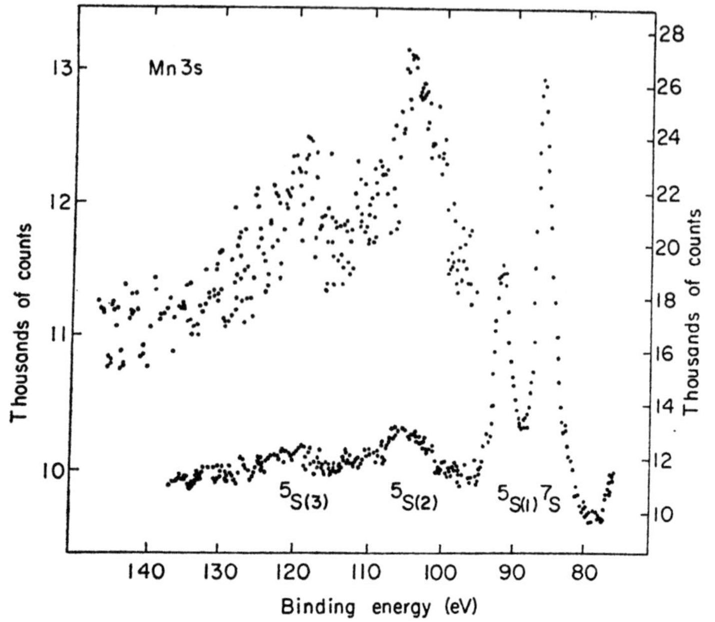


Fig. 19. The 3*s* X-ray photoemission spectrum of $Mn^{2+}$ in $MnF_2$; the right ordinate refers to the complete spectrum, the left ordinate to the expanded part of the spectrum displaying two $^5S$ satellites (see text). Ref. (*46*)

mixing. In the cases of the $Mn^{2+}$ 3*s* and rare-earth 4*s* excitations, however, the situation is not the same. These two levels have the same principle quantum numbers as their respective open, magnetic, valence shells and shakeup mechanisms, similar to those producing the satellites in Figs. 16 and 17, come into play. The calculations of *Bagus* and co-workers (*45*) for the $Mn^{2+}$ $3s^2 3p^6 3d^5 (^6S) \rightarrow 3s^1 3p^6 3d^5$ $(^5S,\ ^7S)$ photoexcitation shed light on this. They show strong mixing of the $3s^2 3p^4 3d^6$ configuration into the $^5S$ final state. This reduces the $^5S$—$^7S$ splitting and introduces satellites. In the spectrum for $MnF_2$ observed by *Kowalcyk et al.* (*46*) and shown in Fig. 19, the satellites are approximately where expected and the ratio of the $^7S$ line intensity to the combined intensities of the $^5S$ lines conforms quite well with Eq. (28). It appears that this class of multiplet structure is rather well understood.

## B. Multiplet Structure in Rare Earth 4*f* Shell Photoemission

Multiplet structure is also observed in the spectra of the open 4*f* shells of the rare earths. *Cox* has provided an explanation (*47*). Consider that the various eigenfunctions are constructed, in Russell-Saunders $J, L, S, M_J'$ coupling, for the $4f^{n-1}$ configuration. Racah showed that a $4f^n$ $[J', L', S', M_J']$ state may be expressed as a sum involving the $4f^{n-1}$ eigenfunctions to which a 4*f* electron is added, namely, neglecting antisymmetrization,

$$\Psi(f^n;\, J'L'S'M_J') = \sum \langle f^{n-1} JLSM_J \cdot f | f^n J'L'S'M_J' \rangle \Psi(f^{n-1};\, JLSM_J) \cdot \phi_f\,,$$

where the $\langle\,\rangle$ are Racah's fractional parentage coefficients and $\phi$, the single $f$ orbital, has suitable quantum numbers. *Cox* pointed out that the probability of excitations to some final $f^{n-1}$ $(J, L, S, M_J)$ state from an initial $f^n(J'L'S'M_J')$ is simply proportional to the square of the coefficient of fractional parentage. In hindsight, this should have been obvious because the coefficients, suitably summed over excitations of each of the $n$ $4f$ electrons in turn, are the overlap integrals of Eq. (7). The spectrum for Sm metal (*36*) is shown in Fig. 20 together with the predicted relative line positions as given by Slater-Condon theory and the Cox predictions for the relative intensities. *Cox* assumed, for the intensity calculations, that the initial state of the rare-earth ion is its Hund's rule $f^n$ ground state. Agreement is excellent. Generally, only a small subset of all the multiplet levels associated with a $4f^{n-1}$ configuration is connected to the ground state and is hence populated in photoemission.

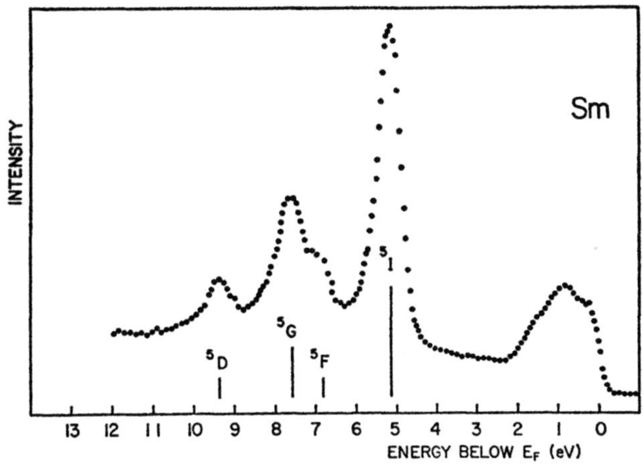


Fig. 20. The high resolution X-ray photoelectron spectrum of the $4f$ shell of Sm metal. With the atom initially in the Hund's Rule $4f^5$ ground state, the vertical marks indicate the relative intensities and positions of accessible final $4f^4$ levels, as predicted by *Cox*. Ref. (*36*)

*Jørgensen* elsewhere in this volume discusses how the Slater parameters appropriate to the multiplet splitting deviate from those associated with the optical multiplet spectra of the $4f^{n-1}$ ions in salts.

As a general rule, the $4f$ spectrum of a rare earth in an alloy or compound is identical to that of the pure metal except for a shift in energy. $SmCo_5$, however, displays behavior not in accord with this rule. Its spectrum and those of Co and Sm metals (*48*) are shown in Fig. 21; Sm metal here was measured at resolution lower than that of the satellite line experiments. In Co and in the compound the conduction bands extend to $-4$ eV. In $SmCo_5$ the intensity of the $4f$ spectrum is relatively low because the ratio of the number of $4f$ to conduction electrons is only 5:48 while it is 5:3 in Sm. Although the quality of the $SmCo_5$ spectrum does not allow a rigorous statement about the matter, it does appear that the structure in the Sm metal $4f$ spectrum is not present in the spectrum of the compound. Perhaps

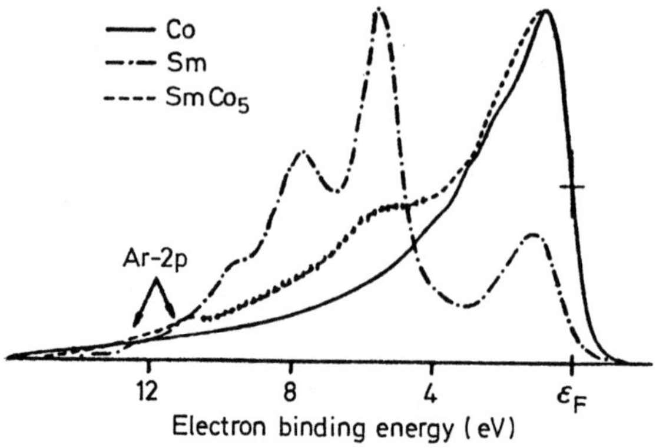


Fig. 21. Conduction and $4f$ electron X-ray photoemission spectra of Sm and Co metals and of $SmCo_5$. Argon sputtering, used to clean samples, resulted in the faint lines seen in the vicinity of 12 eV. Vertical hatching in the 4—10 eV region of the $SmCo_5$ spectrum indicates the uncertainties. Ref. (*48*)

the $4f$ shell in $SmCo_5$ is not in its Hund's rule ground state. Independent data tend to support this supposition: $SmCo_5$ is a ferromagnet having a saturation magnetic moment different from that expected if the $4f$ shell had its normal Hund's Rule moment (*49*). By way of contrast, in $TbCo_5$ the $4f$ photoelectron spectrum and the saturation moment indicate a "normal" $4f$ shell.

Sometimes, more than one $4f$ spectrum is seen to be associated with a single site. Tm in TmTe has a magnetic moment which lies between the values appropriate to divalent and trivalent Tm. In TmSb, on the other hand, the magnetic moment is that characteristic of a $4f$ shell in the trivalent state. Spectra obtained by *Campagna et al.* (50) for the two compounds are plotted in Fig. 22. A single trivalent $4f$ spectrum is observed for TmSb, but TmTe shows a double spectrum with separation $\Delta\mathscr{E} \sim 7$ eV. It appears that the Tm site fluctuates in valence at a rate slow compared with the splitting, $\Delta\mathscr{E}/\hbar$, so that the TmTe spectrum is a "snapshot" of Tm ions in two distinct states of valence[7]). The ratio of intensities of the two valence spectra suggests relative occupation numbers consistent with the observed magnetic moment. Further evidence of the valence fluctuations, the authors suggest, is provided by the splitting of the $5p$ lines, also shown in the figure. It may be noted that there is apparently no evidence in the spectrum for a complementary fluctuation sampled at the Te site.

Core level lines generally display no satellites or shoulders attributable to the differences in local environment among atoms of a given species in a random

---

[7]) If a system is jumping or fluctuating between two states, it is the frequency corresponding to the splitting in the measured spectrum, $\Delta\mathscr{E}/\hbar$ for the case at hand, which determines whether this measurement will show both states individually or a time average of these states. This has been discussed thoroughly in the case of NMR line shapes (*51*). $\Delta\mathscr{E}/\hbar$ is generally much slower than the characteristic time for photoejection of an electron from an atomic site. This latter time in relation to the characteristic times for dynamic screening determines the validity of the sudden approximation.

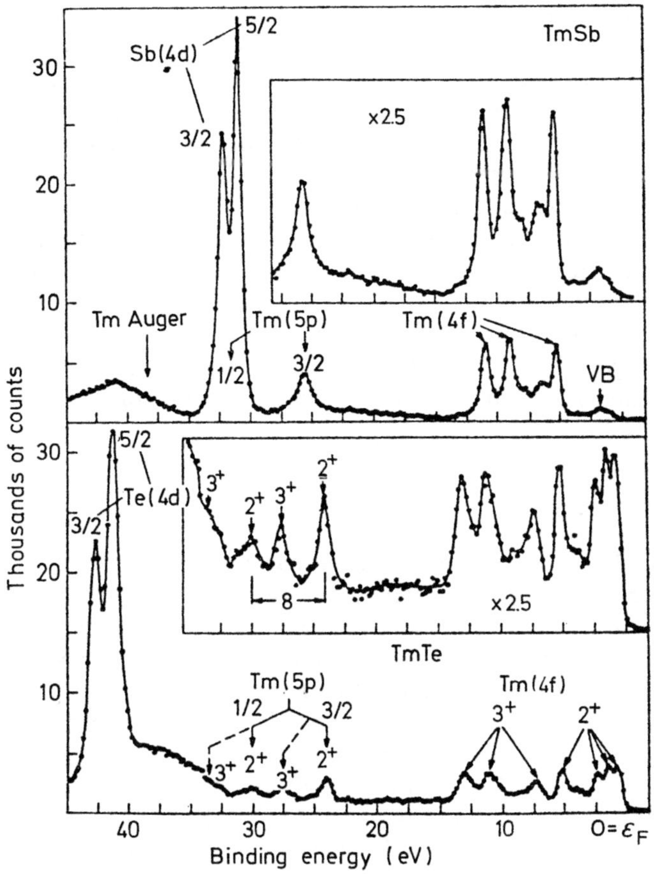


Fig. 22. X-ray photoelectron spectra, extending 44 eV below $\varepsilon_F$, for TmSb and TmTe (see text). Ref. (*50*)

alloy. This is in contrast with results obtained by techniques such as nuclear magnetic resonance and the Mössbauer isomer shift. Of interest, as well, is the difference in local environment between that of an atom in the bulk and one on the clean surface of an otherwise homogeneous sample. Atoms on the surface contribute too small a fraction of the total Mössbauer or NMR signal to be observable. In XPS, however, because of the small depth from which photoelectrons escape, perhaps ten to twenty percent of the photoemission spectrum is derived from the surface layer. One thus might have anticipated that the difference between clean surface and bulk would be detectable by XPS; this has not been the case until now.

On the other hand, the photoelectron spectra of metals are extraordinarily sensitive to the presence of oxygen and other chemical species at the surface. Oxygen core level lines and oxide satellites associated with the metal core lines are readily seen; in fact, with the vacua attained in many spectrometers they are hard to avoid! Preferential oxidations has been observed in measurements of oxide satellites of individual alloy constituents, For example, *Winograd* and his associates found in the Ni—Cr system that Cr is preferentially oxidized (*52*); this led them to postulate a mechanism for the breakdown of nichrome resistors. *Golicheff* (*53*) has observed the preferential oxidation of Sn in the Ni-dilute Sn and Fe-dilute Sn systems. Castle utilized the NiO and CuO shakeup satellites, of the type seen in Fig. 17, for the study of oxidation in Cu—Ni alloys; not surprisingly, Ni is preferentially oxidized in this case (*54*).

On thermodynamic grounds, the composition at the surface of an alloy is sometimes expected to be observably different from that in the interior. Measurements of such differences may be effected from the relative intensities of core level lines of the component elements. In such a study of the Pb—In system, *Berglund* and *Somorjai* (*55*), who used intensities of Auger lines rather than the photoelectron lines themselves, found that Pb tends to concentrate in the alloy surface.

## VII. Surface States

Even though photoelectron spectroscopy has not yet shown evidence for core level shifts corresponding to differences between atoms of a clean surface and atoms in the interior of a solid[8]), both single particle and collective excitations specific to valence electrons of surface atoms have been observed.

XPS studies of surface plasmons of Al have been made by *Barrie* (57) and by *Baer* and *Busch* (*58*). Figure 23 shows Barrie's 2*s* spectrum from a clean Al film and from the film after heavy oxidation. At a 15.2 eV separation from the main peak, and at multiples of this, are peaks associated with bulk plasmons. At 10.7 eV is a small peak, observed with the clean film and not present after oxidation, which originates from excitation of the plasmon characteristic of a clean Al metal surface. *Barrie* concluded that this plasmon energy is in accord with the prediction of *Stern* and *Ferrell* (*59*)

$$\frac{h\omega_p\ (\text{surface})}{h\omega_p\ (\text{bulk})} = \frac{1}{1+\varepsilon_0}, \tag{30}$$

where $\varepsilon_0$ is the dielectric constant of the surface layer.

The detection and measurement through photoemission spectroscopy of the single particle excitation associated with surface valence electron states is made difficult by the accompanying spectrum of the bulk material. When there is a gap in the bulk density of states at $\varepsilon_F$, as in insulators and semiconductors, there is the possibility of observing surface states which lie in the gap. UPS has been used to observe such states in Si (*60*). It is practicable also to search for these states in metals at energies where the densities of bulk valence states are low or relatively structureless. Some suitable candidates for investigation should be the transition metals Ti, Zr, Hf, Cr, Mo, and W, which have low $\varrho(\varepsilon)$ in the vicinity

[8]) Results of some appearance potential experiments have been interpreted as evidence for the existence of such core level differences in the 3*d* transition metals (*56*).

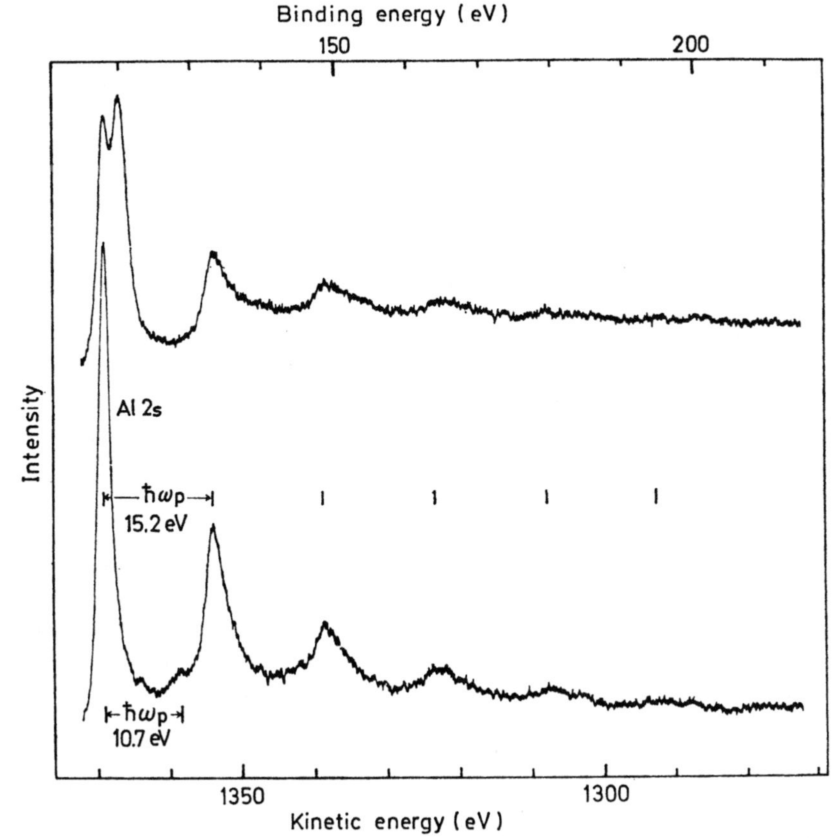


Fig. 23. X-ray photoelectron spectra of the 2s shell of aluminum. Lower curve is for a clean metal surface; the upper curve was observed after exposure to $O_2$ at 1 atmosphere for 30 minutes. Ref. (*57*)

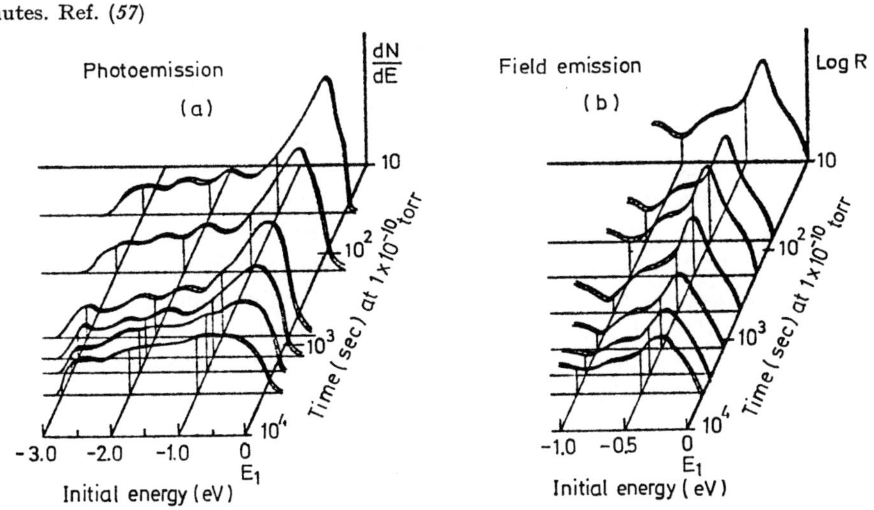


Fig. 24. Ultraviolet photoemission ($h\nu = 7.7$ eV) and analyzed field emission spectra of tungsten as a function of exposure to $H_2$. Log $R$ of the field emission results is related to the local density of states at the surface. Ref. (*61*)

of $\varepsilon_F$. *Waclowski and Plummer* (*61*) have compared field emission and UV photoemission data for tungsten, as shown in Fig. 24. Both sets of spectra exhibit, for the clean metal, a surface state peak at $(\varepsilon_F - 0.4)$ eV which disappears upon adsorption of hydrogen or other gases.

## VIII. In Conclusion

In this chapter we have discussed the various ways in which X-ray photoemission may be employed to explore properties of metals. The experimental methods are rapidly becoming more sophisticated. Improvements in vacuum technique have made possible the investigation of well characterized surfaces and crystal faces; and monochromatization of the regularly used Al and Mg $K$ X-rays has provided narrower photon lines and higher resolution. Development along these lines is continuing and it is reasonable to expect that increasing attention will be given, especially in surface studies, to the dependence of photoemission on photon direction, polarization, and energy as well as on direction and energy of the ejected electrons[9]). Both conventional light sources and synchrotron light (*63*) will be used. The latter, unfortunately available in only a few laboratories, is white, ranging from infrared to X-ray wavelengths; intense; highly collimated; and highly polarized. It is also pulsed, typically in nanosecond bursts; this makes feasible the study of time dependent behavior. Because it provides light at wavelengths not otherwise readily available, studies such as the one by *Eastman* and co-workers (*32*) on Au (Fig. 8) can be made; and some XPS results, also, have recently been obtained with radiation from the Stanford synchrotron facility.

One can expect an increasing tendency to employ, in parallel with XPS, other kinds of experimental data, both for characterization of the samples and their surfaces and for the provision of collateral information. As was seen in the section on core level shifts, these by themselves do not suffice for detailed characterization of the valence electron changes which occur in alloying. In the case of the Au—Sn system, for example, Mössbauer isomer shift data were used to augment the XPS results. Occasional genuine redundancy is to be welcomed.

We have neglected the matter of "chemical" effects on core line widths. Most of the experimentation bearing on this point has involved levels in molecules and insulating solids, and some interesting questions have arisen, such as that of the role that interatomic transitions play in the vacancy filling Auger processes. Recently, *Siegbahn* and co-workers have shown that vibrational effects can make important contributions to the widths observed even for core lines (*64*). Width studies should become more interesting and valuable for furnishing chemical information as higher resolution is attained.

---

[9]) A taste of things to come is given by some toothsome results of this kind recently obtained (*62*) with a single crystal of the layered compound $TaSe_2$. These show strong anisotropies in the valence photoemission, anisotropies which are different for the emissions from the two component elements. Whether this arises from properties of initial state wavefunction or from structural influence on the exiting electrons remains to be seen.

In the realm of theory also, greater demands will be made. As such studies (*37—39*) as those of Cu—Ni (Fig. 13) and Ag—Pd (Fig. 14) have shown, the *d* levels of the two species in transition metal alloys tend to maintain their atomic identities, at least when the levels in the pure components are sufficiently well separated in energy. However, neither calculation nor experiment has been done with refinement sufficient for quantitative testing of a theory, such as the coherent potential approximation, designed to describe the *d* band behavior. In pure metals and intermetallic compounds, band calculations can be compared directly with experiment if transition probabilities and relaxation effects are understood. With care they can be used also in evaluation of the effective interelectronic terms which enter equations such as (18a). Unfortunately, one cannot, by definition, produce a set of "selfconsistent" band calculation results for a matrix of specific valence electron $s^n p^m d^l \ldots$ configurations; thus, direct estimates for $I$ of Eq. (18a) or $F$ of Eq. (18b) cannot be made. However, band calculations for a set of systems can indicate whether or not it is reasonable to factor level shifts into volume and electron count terms, in the manner of Eqs. (18a) and (23). When this cannot be done, one must revert to a more general expression for a level shift, such as Eq. (1).

Core photoelectron ejection produces an atom with a vacancy in some level, a final state which is effectively an impurity imbedded in a surrounding "host". *Herbst et al.* and *Ley et al.*, among others, have provided crude estimates of the extraatomic screening of this impurity; however, many of the methods available for arriving at such estimates are based on perturbation theory, which is not readily applicable to perturbations as large as a vacancy-bearing atomic "impurity". Further from solution is the problem of screening of conduction electron excitations.

In a metal like Ni or Cu there are, below the *d* bands, non-*d* conduction band states, as yet unresolved in XPS spectra; there are heavily hybridized *d*-non *d* states at the bottom of the *d* bands; and there are relatively localized, little hybridized states at the top of the *d* bands. Because the *d* and non-*d* components differ in spatial extent, interatomic screening of the final state hole should vary through the bands. There is little known about how this affects a valence electron XPS spectrum. It is these coexisting complex problems which at times make it difficult to assess agreement or disagreement between predictions of core or valence spectra and the measurements. Does the agreement seen in Fig. 5 indicate that the important physical effects are properly accounted for or is it only the consequence of accidental numerical cancellation?

The parameterization schemes applied here to the core level shifts in Au—Sn and TiC are crude but, we believe, semiquantitative. Their essentials are estimates of Fermi level shifts, of relaxation, of change in atomic volume, and of interelectronic terms calculated, not for free atoms, but for the solid. Refinement is clearly in order.

One can occasionally obtain useful results even when all the factors involved are not understood. For example, relaxation is neglected in the analysis of the TiC core level shifts. The sign of the relaxation effects can be surmised and, with this, it becomes quite clear that the sign and magnitude of the C $\rightarrow$ Ti charge flow deduced would not be affected by inclusion of relaxation. At other times, the quantities of physical interest are not well outside the uncertainties. For example, *Ley et al.* have estimated, from the valence electron spectra, the change in the carbon $2s$—$2p$ count between graphite and diamond (65). It would be useful, if it could be done, to have an independent estimate based on the $1s$ core level shift. Unfortunately, rewriting Eq. (23) for graphite → diamond one discovers that the $\Delta n_s$ term is well within the errors associated with the other terms. Particularly vexacious here, as it often is when dealing with core level shifts, is the uncertainty in $\Delta\varepsilon_F$. This problem can be avoided if one makes use of the shift of one core level relative to that of another, but the cost is the introduction of other bothersome numerical problems. At times it is advantageous to utilize these relative shifts, and at other times it is not.

Since the introduction of X-ray photoemission by the Uppsala group, many questions have been answered and many new ones raised. Widespread and vigorous activity in the field should continue for some time to come.

Note Added in Proof: *Doniach* and *Šunjić* [J. Phys. *C3*, 285 (1970)] have presented a theory for core level line shape assymetry associated with the excitation of low-energy electron-hole pairs in a metal. This skews the otherwise symmetric line towards higher binding energy and such shapes have been observed experimentally [*e.g. P. H. Citrin*, Phys. Rev. B *8*, 5545 (1973); *S. Hüfner, G. K. Wertheim, D. N. E. Buchanan* and *K. W. West*, Phys. Letters *46A*, 420 (1974); and *N. J. Shevchik*, Phys. Rev. Letters *33*, 1336 (1974)]. *Doniach* and *Šunjić* obtained their results within the sudden approximation; and more recently *Gadzuk* and *Šunjić* [to be published] have interpolated the theory between the sudden and the slow, or "adiabatic", limits. Now an electron must be excited 4—5 eV above the Fermi level if it is to overcome the work function barrier and leave the solid. At such energies the electron has a significant velocity, as it escapes the ion core from which it was excited; and, according to *Gadzuk* and *Šunjić*, this places the event in sudden limit. Unless special factors enter, such as resonance effects to slow the escaping electron, it would appear that departures from the sudden limit are unlikely to affect line shape observably.

## IX. References


1. *Siegbahn, K., Nordling, C., Fahlman, A., Nordberg, R., Hamrin, K., Hedman, J., Johansson, G., Bergmark, T., Karlsson, S.-E., Lindgren, I., Lindberg, B.:* ESCA-Atomic, molecular, and solid state structure studied by means of electron spectroscopy. Uppsala: Almqvist & Wiksells 1967.
2. *Siegbahn, K.:* Phil. Trans. Roy. Soc. (London) *A 268*, 33 (1970).
3. *Kanter, H.:* Phys. Rev. B *1*, 522 (1970).
4. *Katrich, G. A., Sarbei, O. G.:* Soviet Phys.-Solid State (English Transl.) *3*, 1181 (1961).
5. *Powell, C. J.:* Surface Science *44*, 29 (1974); another survey in which lower electron energies are emphasized is *I. Lindau* und *W. E. Spicer*. J. Elect. Spectry. *3*, 409 (1974).
6. *Koopmans, T.:* Physica *1*, 104 (1934).
7. *Lang, N. D., Kohn, W.:* Phys. Rev. B *3*, 1215 (1971) and *Lang, N. D.:* Solid State Phys. *28*, 225 (1973); (*Seitz, Turnbull,* and *Ehrenreich*, Eds.). New York: Academic Press.
8. *Hedin, L., Johansson, G.:* J. Phys. *B 2*, 1336 (1969); see also *Watson, R. E.:* Phys. Rev. *118*, 1036 (1960) and *Bagus, P. S., Schaeffer III, H. F.:* J. Chem. Phys. *55*, 1474 (1971).
9. *Lundqvist, B. I.:* Phys. Kondens. Materie *9*, 236 (1969).
10. *Hedin, L., Johansson, A., Lundqvist, B. I., Lundqvist, S., Samathiyakanit, V.:* Arkiv Fysik *39*, 97 (1969). *Manne, R., Åberg, T.:* Chem. Phys. Letters *7*, 282 (1970). — *Meldner, H,. Perez, J. D.:* Phys. Rev. A *4*, 1388 (1971).
11. *Langreth, D. C.:* Phys. Rev. B *1*, 471 (1970). — *Hedin, L., Lundqvist, S.:* Solid State Phys. *23*, 2 (1969); (*Seitz, Turnbull,* and *Ehrenreich*, Eds.). New York: Academic Press. — *Doniach, S.:* In: Computational methods in band theory; (*P. M. Marcus, J. F. Janak,* and *A. R. Williams*, Eds.). New York: Plenum Press 1971. — *Langreth, D. C.:* Nobel Symposium *24*, 210 (1973), Collective properties of physical systems.

12. *Condon, E. U., Shortley, G. H.:* The theory of atomic spectra, Ch. VI. Cambridge: Cambridge Univ. Press 1953.
13. *e.g., Hodges, L., Watson, R. E., Ehrenreich, H.:* Phys. Rev. B *5*, 3953 (1972).
14. *Wigner, E. P., Seitz, F.:* Solid State Phys. *1*, 97 (1955); (*Seitz* and *Turnbull*, Eds.). New York: Academic Press.
15. *Watson, R. E., Ehrenreich, H.:* Comments Solid State Phys. *3*, 109 (1970).
16. *Ducastelle, F., Cyrot-Lackmann, F.:* J. Phys. Chem. Solids *32*, 285 (1971).
17. *Slater, J. C.:* Quantum theory of molecules and solids, Vol. II. New York: McGraw-Hill 1965.
18. *Herbst, J., Lowy, D., Watson, R. E.:* Phys. Rev. B *6*, 1913 (1972).
19. *Ley, L., Kowalczyk, S. P., McFeely, F. R., Pollack, R. A., Shirley, D. A.:* Phys. Rev. B *8*, 2392 (1973).
20. See, in particular, *Hedin* and *Johansson* [Ref. (*8*)]; also see *Shirley, D. A.:* Chem. Phys. Letters *16*, 220 (1972) and Ref. (*1*).
21. *Ehrenreich, H., Gelatt, D., Hodges, L., Watson, R. E.:* (unpublished).
22. *Ramqvist, L., Hamrin, K., Johansson, G., Fahlman, A., Nordling, C.:* J. Phys. Chem. Solids *30*, 1835 (1969). — *Ramqvist, L., Ekstig, B., Källne, E., Noreland, E., Manne, R.:* J. Phys. Chem. Solids *30*, 1849 (1969). — *Ramqvist, L.:* J. Appl. Phys. *42*, 2113 (1971).
23. *Friedman, R. M., Hudis, J., Perlman, M. L., Watson, R. E.:* Phys. Rev. B *8*, 2433 (1973).
24. *e.g., Shirley, D. A.:* Rev. Mod. Phys. *36*, 339 (1964).
25. $IS_1$ is the value of *Charlton, J. S., Harris, I. R.:* Phys. Status Solidi *39*, K1 (1970) with an uncertainty taken larger than their quoted value because of the behavior of the shifts in the Sn-rich region as observed by *Barret, P. H., Grant, R. W., Kaplan, M., Keller, D. A., Shirley, D. A.:* J. Chem. Phys. *39*, 1035 (1963). $IS_2$ is from *Charlton, J. S., Cordey-Hayes, M., Harris, I. R.:* J. Less-Common Met. *20*, 105 (1970). For $IS_3$ an average was taken of the values given by *Charlton, J. S., Cordey-Hayes, M., Harris, I. R.*, and by *Möller, H. S.:* Z. Phys. *212*, 107 (1968); the uncertainty shown covers both values. $IS_4$ is the result of *Charlton, I. S., Harris, I. R.*
26. *Emery, G. T., Perlman, M. L.:* Phys. Rev. B *1*, 3885 (1970).
27. *Rothberg, G. M., Guimar, S., Benczer-Koller, N.:* Phys. Rev. B *1*, 136 (1970).
28. *Watson, R. E., Bennett, L. H.:* Charge density and electronic structure of alloys; (*Bennet* and *Willens*, Eds.). Metallurgical Soc. of AIME, New York, 1974.
29. *e.g., Bennett, L. H., Carter, G. C., Watson, R. E.:* J. Res. Natl. Bur. Std. A *74*, 569 (1970).
30. *Watson, R. E., Hudis, J., Perlman, M. L.:* Phys. Rev. B *4*, 4139 (1971).
31. *Hudis, J., Perlman, M. L., Watson, R. E.:* Am. Inst. Phys., Conf. Proc. *18*, 267 (1973).
32. *Eastman, D. E., Grobman, W. D.:* Phys. Rev. Letters *28*, 1327 (1972) and *Freeouf, J., Erbudak, M., Eastman, D. E.:* Solid State Commun. *13*, 771 (1973).
33. *Dobbyn, R. C., Williams, M. L., Cuthill, J. R., McAlister, A. J.:* Phys. Rev. B *2*, 1563 (1970) and *McAlister, A. J.* (unpublished).
34. *Baer, Y., Hedén, P. F., Hedman, J., Klasson, M., Nordling, C., Siegbahn, K.:* Phys. Scripta *1*, 55 (1970) and Solid State Commun. *8*, 517 (1970).
35. *Fadley, C. S., Shirley, D. A.:* J. Res. Natl. Bur. Std. A *74*, 543 (1970).
36. *Baer, Y., Busch, G.:* Proceedings of the 1974 Namur Conference on Electron Spectroscopy (to be published).
37. *Hüfner, S., Wertheim, G. K., Cohen, R. L., Wernick, J. H.:* Phys. Rev. Letters *28*, 488 (1972).
38. *Hüfner, S., Wertheim, G. K., Wernick, J. H., Melera, A.* Solid State Commun. *11*, 259 (1972); also see *Hedman, J., Klasson, M., Nilsson, R., Nordling, C., Sorokina, M. F., Kljushnikov, O. I., Nemnonov, S. A., Trapeznikov, V. A., Zyryanov, V. G.:* Phys. Scripta *4*, 1 (1971).
39. For example, *Seib, D. H., Spicer, W. E.:* Phys. Rev. Letters *20*, 1441 (1968). — *Norris, C., Nilsson, P. O.:* Solid State Commun. *6*, 649 (1968) or *Norris, C., Myers, H. P.:* J. Phys. F *1*, 62 (1971).
40. *Wertheim, G. K., Rosencwaig, A.:* Phys. Rev. Letters *26*, 1179 (1971).
41. For example, *Krause, M. O., Stevie, F. A., Lewis, L. J., Carlson, T. A., Moddeman, W. E.:* Phys. Letters *31 A*, 81 (1970).
42. *Novakov, T.:* Phys. Rev. B *3*, 2693 (1971)

43. *Fadley, C. S., Shirley, D. A., Freeman, A. J., Bagus, P. S., Mallow, V. J.:* Phys. Rev. Letters *23*, 1397 (1969).
44. *Cohen, R. L., Wertheim, G. K., Rosencwaig, A., Guggenheim. H. J.:* Phys. Rev. B *5*, 1037 (1972). — *Wertheim, G. K., Cohen, R. L., Rosencwaig, A., Guggenheim, H. J.:* In: Electron spectroscopy (*D. A. Shirley*, Ed.) p. 813. Amsterdam: North-Holland 1972.
45. *Bagus, P. S., Freeman, A. J., Sasaki, F.:* Phys. Rev. Letters *30*, 850 (1973).
46. *Kowalczyk, S. P., Ley, L., Pollak, R. A., McFeely, F. R., Shirley, D. A.:* Phys. Rev. B *7*, 4009 (1973).
47. *Cox, P. A.:* Ph. D. thesis, Oxford (1973), unpublished. — *Cox, P. A., Evans, S., Orchard, A. F.:* Chem. Phys. Letters *13*, 386 (1972). — *Cox, P. A., Baer, Y., Jørgensen, C. K.:* Chem. Phys. Letters *22*, 443 (1973).
48. *Cuthill, J. R., McAlister, A. J., Erickson, N. E., Watson, R. E.:* Am. Inst. Phys., Conf. Proc. *18*, 1039 (1973).
49. *Nesbitt, E. A., Wernick, J. H.:* Rare earth permanent magnets. Academic Press New York: 1973.
50. *Campagna, M., Bucher, E., Wertheim, G. K., Buchanan, D. N. E., Longinotti, L. D.:* Phys. Rev. Letters *32*, 885 (1974).
51. *e.g., Carrington, A., McLachlan, A. D.:* Introduction to magnetic resonance, p. 205. New York: Harper and Row 1967.
52. *Winograd, N., Baitinger, W. E., Amy, J. W., Munarin, J. A.:* Science *184*, 565 (1974).
53. *Golicheff, I.:* private communication.
54. *Castle, J. E.:* Nature (PS) *234*, 93 (1971).
55. *Berglund, S., Somorjai, G. A.:* J. Chem. Phys. *59*, 5537 (1973).
56. *Houston, J. E., Park, R. L., Laramore, G. E.:* Phys. Rev. Letters *30*, 846 (1973).
57. *Barrie, A.:* Chem. Phys. Letters *19*, 109 (1973).
58. *Baer, Y., Busch, G.:* Phys. Rev. Letters *30*, 280 (1973).
59. *Stern, E. A., Ferrell, R. A.:* Phys. Rev. *120*, 130 (1960).
60. *Eastman, D. E., Grobman, W. D.:* Phys. Rev. Letters *28*, 1378 (1972). — *Wagner, L. F., Spicer, W. E.:* Phys. Rev. Letters *28*, 1381 (1972). — *Rowe, J. E., Ibach, H.:* Phys. Rev. Letters *32*, 421 (1973).
61. *Waclawski, B. E., Plummer, E. W.:* Phys. Rev. Letters *29*, 783 (1972).
62. *Traum, M. M., Smith, N. V., Di Salvo, F. J.:* Phys. Rev. Letters *32*, 1241 (1974).
63. *Perlman, M. L., Rowe, E. M., Watson, R. E.:* Physics Today, July 1974.
64. *Gelius, U., Svensson, S., Siegbahn, H., Basilier, E., Faxälv, Å., Siegbahn, K.:* Proceedings of the 1974 Namur Conference on Electron Spectroscopy (to be published); for results pertaining more directly to solids see also *Citrin, P. H., Eisenberger, P.,* and *Hamann, D. R.,* Phys. Rev. Lett. *33*, 965 (1974).
65. *McFeely, F. R., Kowalczyk, S. P., Ley, L., Cavell, R. G., Pollak, R. A., Shirley, D. A.:* Phys. Rev. B. *9*, 5268 (1974).

Received July 5, 1974

# Ultraviolet Photoelectron Spectroscopy of Gases Adsorbed on Metal Surfaces

**Alexander M. Bradshaw**

Institut für Physikalische Chemie und Theoretische Chemie der Technischen Universität München, Außenstelle, D-8046 Garching

**Lorenz S. Cederbaum** and **Wolfgang Domcke**

Physik-Department (T 30) der Technischen Universität München, D-8046 Garching

## Table of Contents

## 1. Introduction

The development of gas phase ultraviolet photoelectron spectroscopy (UPS) has proceeded rapidly in the last twelve years. Turner first showed that the process of photoionisation could be used to generate a photoelectron spectrum giving a direct measurement of atomic or molecular ionisation potentials (*1*). The early results of the Oxford group led to the acceptance of two simple guide lines in the interpretation of such spectra. Firstly the electron kinetic energies were considered to satisfy a modified Einstein relation:

$$E = h\nu - I - \Delta E_{\mathrm{vib}} - \Delta E_{\mathrm{rot}}. \quad (1)$$

Secondly it was assumed that Koopmans' theorem holds *i.e.* that the orbital energy computed in an *ab initio* SCF calculation is approximately equal to the ionisation potential, I, of an electron from that orbital. For closed shell systems each orbital energy level is then represented by a band in the photoelectron spectrum, the integrated intensity of which depends on the occupancy of each level, the degeneracy of the ionic state produced by ionisation, and matrix elements. The approximations introduced by the assumption of Koopmans' theorem is discussed fully in several papers [*e.g.* Refs. (*2*) and (*3*)]. A comprehensive review of gas phase UPS is to be found in the recent book by *Eland* (*4*).

About four years ago several groups working in the field of surface chemistry and physics began to think in terms of obtaining similar information about the electronic energy levels of atoms and molecules adsorbed on surfaces. The first paper to show conclusively that such experiments were possible was published in 1971 by *Eastman* and *Cashion* (*5*). They observed changes in the photoelectron spectrum of polycrystalline nickel surfaces as a function of coverage with chemisorbed CO or oxygen, thereby showing that UPS promised much towards solving some of the basic problems of chemisorption. This is not to say that the technique had never been applied to a solid phase: for many years UPS investigations of metals and semi-conductors had been performed to help elucidate problems of band structure. However these early experiments had not sufficiently taken into account the high surface sensitivity of the technique and some of the results must be questioned. It is worthwhile therefore at this point to pause and consider in somewhat more detail why UPS is so surface sensitive and whether we can quantify this factor.

The absorption cross section for vacuum UV photons in a metal is known to be an order of magnitude smaller than the inelastic scattering cross section for electrons of corresponding energy. The surface vs. bulk sensitivity question in the UPS experiment is thus dictated essentially by the mean depth of origin of the photoemitted electron. This mean depth of origin is in turn equivalent to the attenuation length which depends strongly on the inelastic mean free path at the energy concerned[1]). Such electrons in the kinetic energy range 5—50 eV

---

[1]) It is not strictly correct to equate the inelastic mean free path with the attenuation length, as we do so essentially in Fig. 1. One would have to assume in the experiment that as many electrons are scattered elastically into as out of the direction of the analyser slits. Because of the net loss due to back scattering this can never be the case in practice.

above the Fermi level, $E_f$, are subject to two important inelastic scattering mechanisms. The first we may describe as electron-hole-pair creation events, whereby the "hot" electron promotes single electron excitations from states in the filled part of the conduction band to empty states above $E_f$. The second mechanism is scattering from plasmons *i.e.* quantised collective oscillations of the electron gas in the conduction band. As yet few calculations of the extent of these effects are available in the literature. Only for aluminium do we have calculated mean free path data for both plasmon scattering (*6*) and single particle excitation (*7*). Figure 1 shows the calculated mean free path for hot electrons

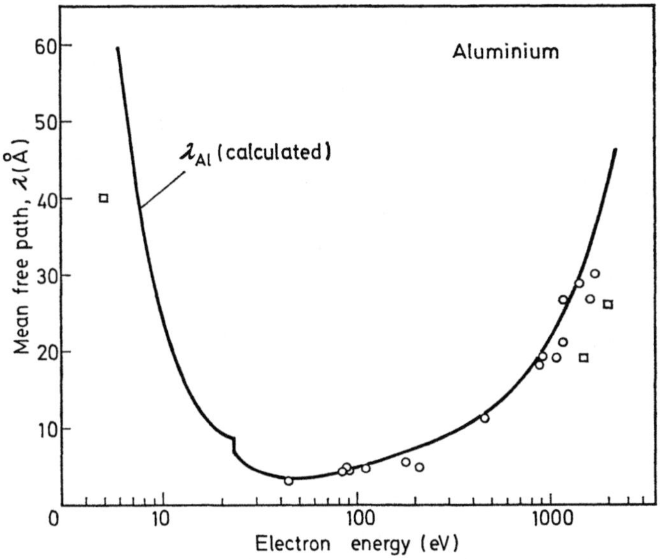


Fig. 1. Mean free path for inelastic scattering, $\lambda$, for aluminium [after Refs. (*6*) and (*7*)]. Experimental points: measured mean attenuation lengths taken from the compilation by *Powell* (*8*)

in aluminium as a function of electron energy between 5 and 2000 eV. Several measured values of the attenuation length are included for comparison, taken from a recent compilation by *Powell* (*8*). The results show that between 5 and 50 eV the mean escape depth for photoemitted electrons in aluminium is expected to lie between 50 and 5 Å. Aluminium is perhaps an extreme case because of its strong intrinsic plasmons: the situation in the transition metals will be slightly less favourable to the surface investigator. In semi-conductors single electron scattering will not be so important because the hot electron must have a minimum energy (relative to $E_f$) of twice the band gap to afford a single particle excitation. At the lowest kinetic energies therefore we expect a much higher mean free path than in metals. This has not however prevented the observation of adsorbate energy levels on silicon surfaces for example (*9*).

In the last three years many more papers on UPS of adsorbed species have appeared in the literature. At a recent Faraday Discussion *(10)* on photo-effects in adsorbed layers no less than nine papers dealt with this topic. It is therefore appropriate at the present time to review briefly some of the work that has been done already, to consider which way the subject is leading and to pick out what might be important for experimentalists in the future. In this sense the present article does not set out to be a comprehensive review, but rather a progress report.

We must first however define exactly our problem. What do we aim to achieve by doing UPS of adsorbates? In simple terms we are interested in the electronic structure of the metal-adsorbate-complex. The coupling of the adsorbate to the surface converts the originally discrete levels which are co-incident with the continuum of metal states into virtual levels or resonances *(11, 12)*. The occupancy of these resonances is determined by their position on the energy scale relative to $E_f$. It has been proposed that the concept of a "surface molecule" *(13—15)* can be used to describe the adsorbate-adsorbent interaction. Several questions thus arise. Is the concept of the surface molecule valid? What are the rules for its formation? Is it weakly or strongly bound together? To what extent is it necessary to take many body effects into consideration? A further problem arises in the interpretation of the photoemission process in the adsorbed layer. Can the problem be considered in terms of photoionisation of a surface molecule or do we need to picture it as photoejection from an N-electron ground state of the metal-adsorbate-vacuum complex? For a quantitative understanding of UPS of adsorbates we obviously need in the first instance a satisfactory theory of chemisorption and then a theory which links this theory in turn with photoeffects in adsorbed layers. To what extent progress has already made along these lines we discuss in Section 3.

It might well be asked whether it is worthwhile carrying on with UPS of adsorbates in the absence of such a quantifying theory. How much useful information can we obtain without a complete theory? In Section 4 we look at several examples of chemisorption systems, which have been investigated with UPS, concentrating mainly on CO as adsorbate. In particular the CO/Ni system is discussed in terms of less sophisticated theories of chemisorption, based on the assumption of a simplistic single particle photoionisation process which we saw above has been used with success in gas-phase work. We then discuss briefly the chemisorption of oxygen and of hydrocarbons, before going on to review progress on the measurement of angular distributions, which we shall see in Section 3 could play an important role in maximising the information content in UP-spectra.

## 2. Experimental Aspects

We do not intend under this heading to give a complete account of the experimental problems and pitfalls encountered in the UPS of adsorbates. Nonetheless some of the existing experimental arrangements will be briefly described and future developments will be discussed. For convenience the chapter is divided into experimental aspects associated with the radiation source, the vacuum environment and surface conditions, the analyser and finally the detection system.

## 2.1. Sources

The properties which we require from an ideal source in UPS is good monochromaticity, defined by a line width better than 100 meV, high intensity and continuous wavelength selectivity. Unfortunately only synchrotron radiation together with a suitable monochromator offers us all these features. The rare gas discharge lamps or, alternatively, monochromators with continuum or many-line sources have generally been used hitherto, but have certain disadvantages.

The rare gas discharge lamps, operated with helium or neon, offer a line spectrum whereby one line, normally designated the *raie ultime*, is the strongest. Thus the *raie ultime* in the HeI spectrum occurs at 584.334 Å or 21.22 eV. (The I implies that the radiation is emitted by the neutral atom, II implies a singly charged ion, III a doubly-charged ion etc.) The width of the lines depends upon the natural widths of the energy levels involved and upon external conditions such as temperature and pressure. Typically they are of the order of few milli-electron-volts. The excited species are normally created in a glow discharge, often called a cold cathode discharge. Several commercial companies now market such discharge lamps, although many workers have built their own with considerable degrees of success. [For a survey of various lamp designs, the reader is referred to the book by *Samson* (*16*)]. The most widely used line, also in gas phase work, is the HeI line at 21.2 eV. The *raie ultime* from ionised helium (HeII) at 40.81 eV is almost always present as well. In order to produce a higher proportion of HeII it is necessary to lower the gas pressure and to increase the current density. Unfortunately the normal result of such an operation is that the discharge goes out. It remains a problem to design a discharge lamp which gives as much if not more HeII than HeI, without recourse to metal foil filters, which have to be extremely thin in order to avoid too much loss of intensity.

Many-line or continuum sources in conjunction with monochromators have been used, but are generally low in intensity, not continuous and restricted anyway to higher wavelengths.

The most promising source for the future in UPS studies is undoubtedly synchrotron radiation (*17—19*). With a suitable monochromator synchrotron radiation offers us high intensities and a continuous range of energy from the microwave to the X-ray region. The classical mechanism by which electrons lose energy in an accelerating field is by the production of electromagnetic radiation. In a synchrotron or storage ring the accelerating field is provided by magnetically constraining the electrons to travel around a closed path. The consequent centripetal acceleration results in the emission of synchrotron radiation. Relativistic effects result in the radiation being emitted in a very narrow angular distribution about a tangent to the electron orbit. The intensity increases with increasing energy up to a maximum somewhere in the soft X-ray region, depending on the energy and curvature of the electrons. The intensity of existing synchrotron sources seems to be sufficient for photoelectron spectroscopy; the new generation of storage rings coming into use (*e.g.* DORIS in Hamburg) will certainly be adequate for most experimental purposes. (The storage ring in general has a very small source size which, together with the low divergence, makes for a high source brightness). *Eastman et al.* (*20*) have already published UPS results from certain semi-conductors using storage ring radiation at Cambridge, Mass.

To sum up, synchrotron or storage ring radiation offers many advantages, in particular the continuous energy distribution and in addition almost 100% polarisation in the plane of the ring. To its disadvantage is the fact that there are few such machines available and the researcher must take his apparatus to the facility — assuming that he can get place to work there.

## 2.2. Preparation and Maintenance of Clean Surfaces

The preparation of clean surfaces in uhv has become by now a standard procedure. Several texts deal with this problem [see for example Refs. (*21*) and (*22*)]. Single crystal surfaces are generally cleaned by chemical means (*e.g.* oxidation and subsequent flashing in the case of tungsten) or by argon ion sputtering (*23*) or by a combination of both (*24*).

Particularly good vacuum conditions in UPS of adsorbates are required to *maintain* clean surfaces, as discussed by *Bradshaw* and *Menzel* (*25*). With any form of analyser increased scan times or an increased number of scans improve the quality of the spectrum from the point of view of signal-to-noise ratio. Unfortunately long scan times increase the danger of contamination of the surface. This is particularly true of clean surfaces or of surfaces only partly covered with adsorbate, where the sticking coefficient for residual, adsorbable gases is high. Thus there is a third parameter to consider when we make the usual compromise between resolution and sensitivity. In general it seems better to aim for adequate resolution, so that the width of the adsorbate levels is not noticeably reduced by reducing the slit width of the analyser, or in the case of the retarding field analyser, the modulation voltage. (In the case of certain metals such as tungsten the structure in the valence band may contain sharper features, which can then be used as a better criterion). When the resulting sensitivity is so poor that an acceptable signal-to-noise ratio is only obtained through long scan times, then it is better in the case of metals with high sticking coefficients for reactive gases to make one measurement at a particular coverage, clean the crystal and then make a further measurement at another coverage and so on. *Bradshaw* and *Menzel* (*25*) have found, even with their background pressure of $2–3 \times 10^{-11}$ Torr of reactive gases, that the maximum allowable scan time for clean or partially covered surfaces is 15—20 minutes.

Because discharge lamps are invariably windowless and differentially pumped, the partial pressure of helium in the vacuum chamber is likely to rise considerably during the experiment. A mass spectrometer for residual gas analysis is therefore absolutely necessary to check the purity of the helium.

## 2.3. Analysers

The simplest form of electron energy analyser is the integrating retarding field system shown schematically with suitable circuitry in Fig. 2. It operates on the principle of allowing only those electrons through to the collector, which have energies higher than the retarding voltage applied to the two middle grids. The

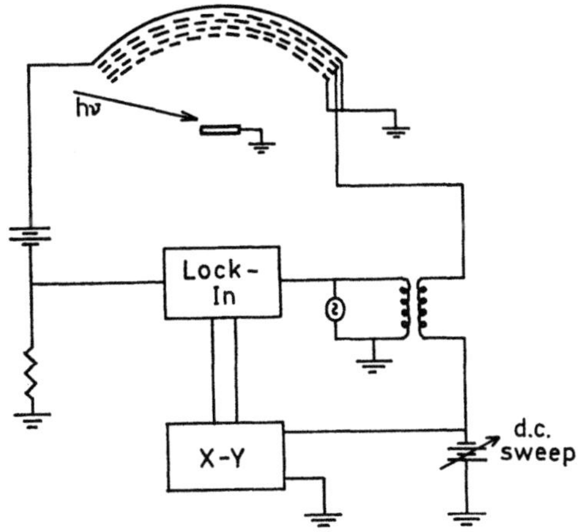


Fig. 2. Typical retarding field system for UPS measurements

spectrum is obtained by scanning the retarding voltage. Because such an operation produces an integrated form of the desired spectrum, it is used in combination with a well-known modulation technique (*26*) to produce the first derivative, or real spectrum. The reason why the form of the retarding field analyser as depicted in Fig. 2 has found favour among surface physicists is that essentially the same equipment has been used for several years in LEED and Auger investigations of surfaces. Because the angular dependence of the emitted photoelectrons is deliberately ignored, it is better perhaps to integrate over as large a solid angle as possible and to use hemispherical grids. Resolution and sensitivity depend on the maintenance of a field free space between the crystal and first grid, on the balancing out of capacitative coupling effects between retarding grid and collector and on the size of the voltage modulation. In practice it is difficult to get a resolution much better than 0.2 eV.

Higher resolution is offered by dispersive analysers of which the main types are the 127° (*27*), the hemispherical (*28*), and the cylindrical mirror analyser (*29*). The various types will not be discussed here and the reader is referred to the useful summaries by *Eland* (*4*) and *Sevier* (*30*). At least two commercial companies offer uhv photoelectron spectrometers for surface work combining hemispherical analysers with discharge lamps. *Bradshaw* and *Menzel* describe (*25*) a system for surface studies, whereby other techniques can be combined with photoelectron spectroscopy (see Fig. 3).

In all systems hitherto the dispersive analysers have been stationary. Although it may be possible to move the sample this automatically changes the angle of incidence of the exciting radiation. For the angular distribution measurements, which appear to be necessary in the future, new purpose-built systems will be required. Measurements at fixed exit angles may not be exactly obsolete but, as we shall see in Section 3 below, leave a lot to be desired. Two instruments based on the retarding field principle and suitable for angular distribution measurements have been described however in the literature. The first due to *Smith* and co-

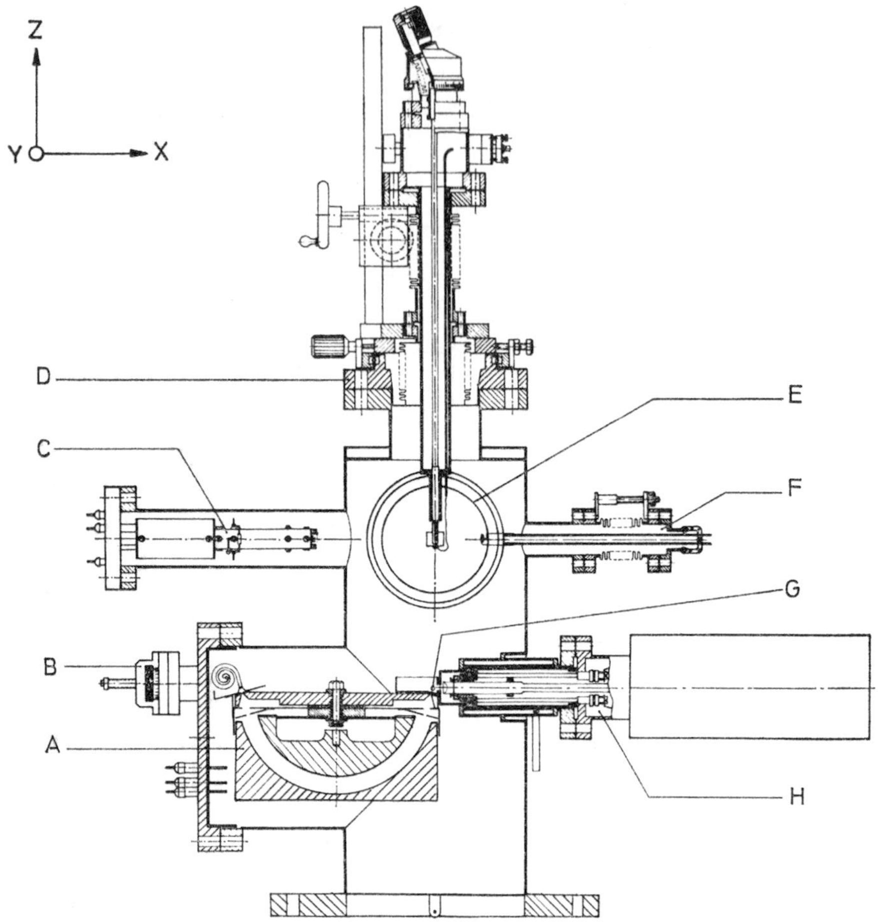


Fig. 3. Cross-section of the combined UPS-XPS-LEED system described by *Bradshaw* and *Menzel* (*25*). *A* - Hemispherical electron energy analyser; *B* - Slit-change mechanism; C - Quadrupole mass spectrometer; *D* - Crystal manipulator; *E* - LEED optics; *F* - Assembly for electron bombardment of the sample; replaceable by an argon ion gun; *G* - Rare gas resonance lamp with axis perpendicular to the plane of the cross-section; *H* - X-ray source. [Reproduced with permission from Ber. Bunsenges. *78*, 1140 (1974)]

workers at Bell Laboratories has as yet only been used to measure bulk angular distributions from GaAs (*31*) and $TaSe_2$ (*32*), the latter being a layer compound, resembling perhaps in some respects an adsorption system. This group is currently performing angular distribution measurements on the $(2 \times 1)$ reconstructed (111) face of silicon (*33*). The apparatus consists simply of a hemisphere maintaining a field free region between itself and the sample and containing a slit behind which a small retarding field analyser moves (see Fig. 4). This arrangement scans the polar angle and the sample is then rotated to scan the azimuthal angle. The retarding potential is applied to the second of the two lens elements and using the modulation technique a resolution of about $\sim$0,3 eV seems to be obtain-

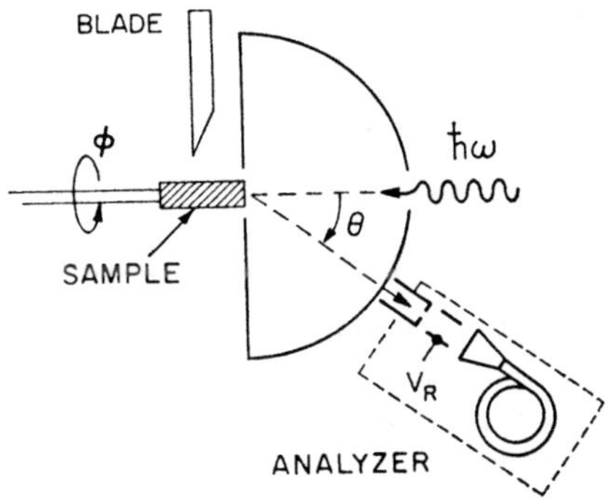


Fig. 4. Schematic diagram of the experimental apparatus of Smith and co-workers showing the position of the GaAs crystal, cleaving blade and photoelectron energy analyser. [Reproduced with permission from Phys. Rev. Letters *31*, 1247 (1973)]

ed. The second published angular distribution is due to *Waclawski et al.* (*34*) and consists essentially of a three grid, modulated retarding field analyser (see Fig. 5). Electrons photoemitted from the sample pass through the analyser grids and strike a 76 mm diameter channel plate. A flat fluorescent screen positioned behind the channel plate provides a visual display of the amplified photoelectron current. A spot photometer is used to measure the intensity at a small, selected area of the screen. The sampled area at the centre of the screen corresponds to a solid angle of $\sim 1.2 \times 10^{-2}$ steradians ($\sim 3.5°$ half-angle). Because of distortion effects this decreases to $\sim 4 \times 10^{-3}$ steradians ($\sim 2°$ half-angle) towards the edges of the screen. The a.c. modulated component of the screen intensity within the sample area is converted by the spot photometer to an a.c. voltage, which goes to a lock-in amplifier. Two modes of operation are possible: one can either measure a complete spectrum at a fixed emission angle or one can look at a particular feature in the energy distribution curve over a range of emission angles by setting at fixed retarding potential. Some of the first results from this system are discussed below in Section 4.

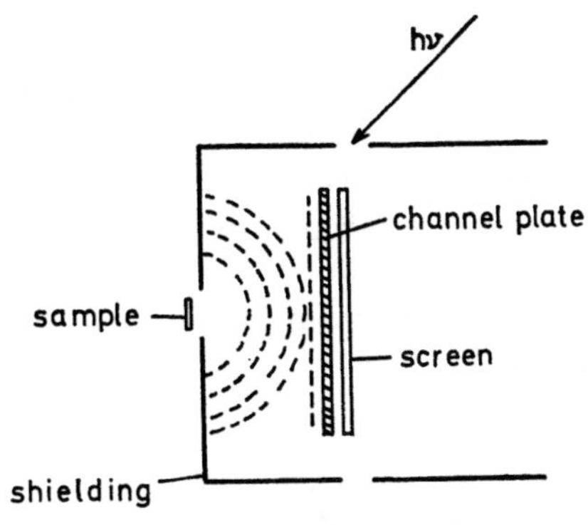


Fig. 5. Schematic diagram of a retarding field analyser system using a channel plate. [After *Waclawski et al.* (*34*) and other groups.]

### 2.4. The Detection System

In the retarding field system we have seen that the most satisfactory form of measurement uses potential modulation in conjunction with a lock-in amplifier. If the latter is then simply combined with a X-y recorder, then the photoelectron energy distribution is simultaneously graphically displayed. If we wish to measure the so-called difference spectrum[2]), we must use an analogue-digital converter and feed the data to a cumputer or signal averager.

A dispersive analyser used in UPS gives typically currents ranging from $10^{-13}$ to $10^{-18}$ Å, and a simple collector plus amplifier can no longer be used. Under these conditions counting techniques prove satisfactory. An electron multiplier is used to convert the single electrons into pulses, which can be counted digitally or converted into analogue form with a ratemeter. The counting can be performed either with a single scaling device, which means that the data must be stored on cards or tape for later processing, or by a multi-scaler. The most convenient multi-scaling system is undoubtedly a multi-channel analyser linked "on line" to a computer. A good second best is the "signal averager", a simple type of multi-channel analyser, containing in addition facilities for adding and subtracting portions of the memory.

In general it is better to make many scans of the spectrum in the time available, rather than one single scan. Simple considerations show that only when the noise is "white" is the signal-to-noise ratio the same in both modes. In practice, white noise is never observed.

## 3. Aspects of the Theory

The spectra observed in gas phase UPS reveal directly the ionisation potentials and the vibrational structure due to ionisation. The matrix element for the ionisation process is, to a good approximation, usually constant over the width of the corresponding band and therefore, for each band, the measured spectrum is closely related to the density of occupied states of the molecule. Since the latter quantity is also accessible theoretically, a direct comparison between measured and calculated spectra is possible (*35*).

Unfortunately the situation is considerably more involved if photoemission from adsorbates is considered. There is always a contribution to the photoelectron current from the underlying substrate which, by itself, usually exhibits considerable structure both in the energy as well as in the angular distribution. This fact is accounted for in current experimental work by taking the so-called "difference spectra", referred to above. These difference spectra are considered by many experimentalists to contain the desired information about the electronic structure of the surface adsorbate complex.

---

[2]) The difference spectrum is the result of subtracting the clean surface spectrum from the spectrum of the adsorbate-covered surface. There is no agreement among workers in this field as to whether the difference spectrum or the spectrum of the adsorbate-covered surface is the one most easily interpreted.

In chemisorption theories (*36—38*) the electronic structure of the surface-adsorbate complex is usually described in terms of the density of states at the adsorbate, given by the imaginary part of the Green's function at the adsorbate. The purpose of the present section is to elucidate to what extent and on what premises the observed energy distribution in the difference spectrum reflects directly the density of states at the adsorbate. Since the angular distribution of the photoelectron current can also be measured and such spectra will be available in the near future, the question arises as to whether the observed angular anisotropies may provide us with information about the geometry of the adsorbate bonding site and the bonding orbitals.

The first step towards the formulation of a theory of photoemission from adsorbate-covered surfaces was due to *Penn* (*39*). Working within a simple model of chemisorption and invoking certain simplifying approximations he was able to express the excess current $\Delta j(\omega)$, produced by the adsorbate at energy $\omega$, in terms of the real and imaginary parts of the adsorbate Green's function. The starting point for the derivation of Penn's result is the familiar Golden Rule, giving for the total photoemission current per unit energy at energy $\omega$,

$$j(\omega) = 2\pi \sum_F |\langle F|H'|I\rangle|^2 \, \delta\,(E_F - E_I - \omega_0)\, \delta\,(\omega - \varepsilon_f) \tag{2}$$

where $|I\rangle$ and $|F\rangle$ are the initial and final $N$-electron states, and $E_I$ and $E_F$ the corresponding energies. $\omega_0$ denotes the energy of the incoming photons and $\varepsilon_f$ the energy of the photoemitted electrons. The operator $H'$ describes the coupling of the electrons to the radiation field

$$H' = \frac{e}{mc} \sum_i \boldsymbol{A} \cdot \boldsymbol{p}_i. \tag{3}$$

Assuming the photoemitted electron in the continuum state $|f\rangle$ to be no longer correlated with the other electrons, Eq. (2) can be rewritten in the form

$$j(\omega) = 2\pi f(\omega - \omega_0) \sum_{ii'f} \langle f|\tau|i\rangle \langle i'|\tau|f\rangle$$
$$\frac{1}{\pi}\, \mathrm{Im}\, G_{ii'}(\omega - \omega_0)\, \delta\,(\omega - \varepsilon_f)\,, \tag{4}$$

where $\tau = \frac{e}{mc} \boldsymbol{A} \cdot \boldsymbol{p}$. The summation is over a complete set of single particle states $|i\rangle$ of the metal plus adsorbate; $G_{ii'}(\omega - \omega_0)$ denotes the corresponding elements of the Green's function and $f(\omega)$ is the Fermi factor. $f(\omega) = 1$ if $\omega \neq E_f$, $f(\omega) = 0$ if $\omega \neq E_f$, $E_f$ being the Fermi energy of metal plus adsorbate. Manybody effects which are present in the initial and final state wave functions $|I\rangle$ and $|F\rangle$ in Eq. (2) are reflected in the Green's function $G_{ii'}(\omega - \omega_0)$.

A simple and widely used model to describe chemisorption is the Anderson model. The Anderson Hamiltonian (*40*) reads

$$H = \sum_k \varepsilon_k n_{k\sigma} + \varepsilon_a n_{a\sigma} + U n_{a\uparrow} n_{a\downarrow} + \sum_{k\sigma} (V_{ak} c^+_{a\sigma} c_{k\sigma} + V^*_{ak} c^+_{k\sigma} c_{a\sigma}), \tag{5}$$

where $n_{k\sigma} = c^+_{k\sigma} c_{k\sigma}$, $n_{a\sigma} = c^+_{a\sigma} c_{a\sigma}$. The states $|k\rangle$ represent the continuum of single particle states in the metal and the subscript $a$ refers to the adatom orbital $|a\rangle$. A single adatom orbital is assumed for simplicity. $c^+_{k\sigma}$ and $c^+_{a\sigma}$ are creation operators for an electron of spin $\sigma$ in the orbitals $|k\rangle$ and $|a\rangle$, respectively. The parameter $U$ describes the Coulomb repulsion of two electrons at the adsorbate. The coupling between adsorbate and metal is provided by the hopping matrix elements $V_{ak}$. Identifying the set of single particle states $\{|i\rangle\}$ in Eq. (4) with the set $\{|k\rangle, |a\rangle\}$ of the Anderson model and using the Hamiltonian Eq. (5), one obtains for the difference in photocurrent due to the presence of the adsorbate, $\Delta j(\omega)$,

$$\Delta j(\omega) = \alpha(\omega) \operatorname{Im} G_{aa}(\omega - \omega_0) + \beta(\omega) \operatorname{Re} G_{aa}(\omega - \omega_0), \tag{6}$$

where the coefficients $\alpha(\omega)$ and $\beta(\omega)$ depend on the electronic structure of the substrate as well as on the hopping matrix elements $V_{ak}$ and are given by *Penn* (*39*). In the case of no coupling between the adsorbate and the metal, *i.e.* $V_{ak} = 0$, the coefficients are simply $\beta(\omega) = 0$ and $\alpha(\omega) = 2f(\omega - \omega_0) \times \sum_f \delta(\omega - \varepsilon_f) |\langle f|\tau|a\rangle|^2$. Therefore, in this particular case, $\Delta j(\omega)$ provides a direct measure of the density of occupied states at the adsorbate, $1/\pi \operatorname{Im} G_{aa}(\omega - \omega_0)$, if there is no pronounced structure in the density of final states and if the variation of the matrix element $\langle f|\tau|a\rangle$ with the final state energy can be neglected over the width of the adsorbate density of states function. The angular distribution of $\Delta j(\omega)$ is determined by the matrix element $\langle f|\tau|a\rangle$ and reflects the symmetry of the adsorbate orbital.

Due to the interaction of the adsorbate orbital with the continuum of metal states, however, there is an interference of the photoelectron current from the adsorbate with that originating from the substrate, leading to the appearance of the second term in Eq. (6). As shown by *Penn* (*39*) a peak present in $\operatorname{Im} G_{aa}(\omega - \omega_0)$ may be distored and shifted by the contribution of $\operatorname{Re} G_{aa}(\omega - \omega_0)$. If, for example, $\operatorname{Im} G_{aa}(\omega - \omega_0)$ can be represented by a Lorentzian, $\operatorname{Re} G_{aa}(\omega - \omega_0)$ exhibits an antiresonant behaviour and, depending on the coefficients $\alpha(\omega)$ and $\beta(\omega)$, the line shape in $\Delta j(\omega)$ becomes asymmetric. This effect has been termed the "anti-resonance effect" (*39*). Furthermore it is difficult to assess the energy dependence of the coefficients $\alpha(\omega)$ and $\beta(\omega)$. A strong variation of these coefficients with energy will further obscure the structure due to $\operatorname{Im} G_{aa}(\omega - \omega_0)$ in the observed difference spectrum $\Delta j(\omega)$. For these reasons it is not clear at all to what extent the measured difference spectra bear resemblance to the quantity of interest, the density of states at the adsorbate.

A more favourable situation should pertain if the adsorbate level is located below the conduction bands of the metal. In this case the density of metal states vanishes in the region of the resonance level and the measured photoelectron current should directly reflect, apart from matrix element variations, the additional density of states due to the presence of the adsorbate. Such a situation is indeed encountered in some adsorption systems, *e.g.* in CO adsorption on transition metals, where some adsorbate induced levels are clearly separated at least from the metal *d*-band. Hopefully for these systems a straightforward interpretation of the observed spectra in terms of the density of states at the adsorbate is possible.

Some supplementary remarks to the theory of Penn might be appropriate here. There are additional effects which are of relevance if a more quantitative theory of the photoemission process from an adsorbate-covered surface is envisaged. The first point is that the Anderson model as applied to chemisorption is a clearly oversimplified model to describe real metal-adsorbate systems. Besides overlap effects due to the nonorthogonality of the states $|k\rangle$ and $|a\rangle$, there are several interaction effects which are neglected in the Hamiltonian, Eq.(5). The adsorbed atom, for instance, may act as a scattering centre for the metal electrons and thus modify the Bloch wave functions characteristic of the free substrate. This can be accounted for by adding a term

$$\sum_{kk'} \tilde{V}^{kk'} a_k^+ a_{k'} \tag{7}$$

to the Hamiltonian Eq. (5), the $\tilde{V}^{kk'}$ being the matrix elements of the adsorbate potential in the basis of metal states $|k\rangle$. An interaction of this type is expected to cause a redistribution of electronic charge at the surface due to the perturbation of the substrate by the adsorbate potential. There is experimental evidence that such redistribution effects can occur, an example being the disappearance through adsorption of surface states present at the clean surface (*41*, *42*).

The derivation of Eq. (6) as outlined above can be extended to include the interaction term Eq. (7) and one may formally write for the photoelectron current

$$\begin{aligned} j(\omega) = 2f(\omega - \omega_0) \operatorname{Im} \sum_f \delta(\omega - \varepsilon_f) [\langle f|\tau g \tau|f\rangle + \\ \langle f|\tau(1 + g\tilde{V})|a\rangle G_{aa}(\omega - \omega_0) \langle a|(1 + \tilde{V}g)\tau|f\rangle], \end{aligned} \tag{8}$$

where $g$ denotes the Greenian operator in the subspace of metal states, with matrix elements

$$\begin{aligned} g_{kk'} &= g_k \delta_{kk'} + \sum_{k''} g_k \tilde{V}^{kk''} g_{k''k'}, \\ g_k &= \frac{1}{\omega - \omega_0 - \varepsilon_k}. \end{aligned} \tag{9}$$

Defining furthermore a state $|b\rangle$ by

$$|b\rangle = \frac{1}{V} \sum_k V^{ak} |k\rangle, \tag{10}$$

where $V = (\sum_k |V^{ak}|^2)^{\frac{1}{2}}$ is a measure of the adsorbate substrate coupling strength, Eq. (8) reads

$$j(\omega) = 2f(\omega - \omega_0)\ \mathrm{Im} \sum_f \delta(\omega - \varepsilon_f)[\langle f|\tau g\tau|f\rangle + (\langle f|\tau|a\rangle + V\langle f|\tau g|b\rangle)\, G_{aa}(\omega - \omega_0) \times (\langle a|\tau|f\rangle + V\langle b|g\tau|f\rangle)]. \tag{11}$$

The state $|b\rangle$ represents the particular linear combination of metal states which actually participates in the chemisorption bond and may be considered to be localised at the surface in the vicinity of the adatom. It follows from Eq. (11) that the energy distribution of the photoemission current depends not only on both the real and imaginary parts of $G_{aa}(\omega - \omega_0)$ but also on the properties of the propagator $g(\omega - \omega_0)$. In the surface molecule limit *(13–15)*

$$\langle f|\tau g|b\rangle \approx \langle f|\tau|b\rangle g_b(\omega - \omega_0), \tag{12}$$

and it is clear that sharp structure in Im $g_b(\omega - \omega_0)$ may largely obscure the structure introduced by $\mathrm{Im}\, G_{aa}(\omega - \omega_0)$, if the latter is a more smoothly varying function of energy. Thus it depends on the relative half-widths of $\mathrm{Im}\, G_{aa}(\omega - \omega_0)$ and $\mathrm{Im}\, g_b(\omega - \omega_0)$ as to whether the energy distribution of the photoemission current reflects the properties of the adsorbate or a structure more typical of the substrate. Due to the scattering of the metal electrons by the adatom as described by Eq. (9), $\mathrm{Im}\, g_b(\omega - \omega_0)$ is not simply the density of states at the clean surface.

A further complication arises due to final state scattering effects. As will be discussed below, it is important to consider the Bloch character of the final state wave function $|f\rangle$, *i.e.* to take into account the scattering of the outgoing photoelectrons by the periodic array of metal atoms *(53)*. In addition, the adatom itself acts as a scattering centre for the photoemitted electrons and modifies the final state wave function. This latter effect has not yet been incorporated into the theory of the photoemission process. It should be noted, however, that the effect of this scattering, as well as that of the scattering of the metal electrons by the adatom, cannot be eliminated by simply lowering the adsorbate coverage. Both effects are due to the local perturbation of the surface by the adsorbed species and are, as the adsorbate induced photoemission current itself, proportional to the coverage. On the other hand, perturbations due to the interaction of two or more adatoms or to multiple scattering from different adsorbates are proportional to the square of the coverage. It is only the latter effects which become negligible at low coverages.

The discussion so far has concentrated on the energy distribution of the photoemission current. Indeed the experimental devices which are in use at present measure, as a function of kinetic energy, the current collected either over a small fixed solid angle or over a larger solid angle centered around the surface normal. Considerably more information could be obtained if the angular distribution of the photoelectrons were measured. The angular distribution of photoelectrons has been investigated both experimentally and theoretically in the case of atoms and molecules in the gas phase (*43—45*) and in the case of photoemission from crystalline materials (*31—33, 46*); see also Section 2.3 above. In photoemission from crystals the diffraction of the outgoing electrons is considered to cause pronounced angular anisotropies (*46*). The observed angular distributions are therefore in part due to final state effects. For atoms and molecules in the gas phase and unpolarised photons an angular distribution of the form

$$I(\Theta) = 1 + \frac{1}{2}\,\beta\left(\frac{3}{2}\sin^2\Theta - 1\right), \tag{13}$$

$\Theta$ being the angle between the directions of the incoming photon and the outgoing electron, is generally observed. The "angular parameter" $\beta$ depends on the atomic or molecular orbital from which photoionisation takes place. For atoms and molecules in the gas phase the angular anisotropy is considered to be mainly an initial state effect (*45*). Most of the information about the initial state, however, is lost due to the random orientation of the molecules. It is clear that the situation is more favourable in the case of atoms or molecules adsorbed on a single crystal surface, since there is now a fixed orientation of the orbitals of the adsorbed species with respect to the surface. Thus one can expect the angular distribution of the photoemission current from adsorbates to provide detailed information about the geometry of the bonding orbitals. However, due to the periodicity of the underlying substrate, final state induced angular anisotropies will also be present and therefore the theoretical calculation of the angular distribution amounts to a formidable problem.

The dependence of the angular distribution of the adsorbate - induced photoelectron current on the initial state wave function has been recently calculated by *Gadzuk* (*47—49*). He has considered adsorption of atoms on transition metals, assuming the interaction of the adatom with the $d$-type orbitals of the substrate to be responsible for the chemisorption. To avoid the difficulties associated with Penn's antiresonance effect he assumed the adsorbate level to be non-degenerate with the metal $d$-band. Neglecting the variation of the optical matrix element over the width of the adsorbate level the angular distribution is in this case given by

$$\frac{dj}{d\Omega} \sim |\langle f|\tau|\tilde{a}\rangle|^2 \,\mathrm{Im}\, G_{aa}(\omega - \omega_0)\,\varrho_f(\omega) \tag{14}$$

where $|\tilde{a}\rangle$ denotes the one-particle wave function of the particular resonance level in question and $\varrho_f(\omega)$ is the density of final states. In the surface molecule model $|\tilde{a}\rangle$ is represented by a linear combination of the adatom orbital and the orbitals of the nearest neighbour substrate atoms. If the final state wave function $|f\rangle$ is approximated by a plane wave $e^{i\boldsymbol{k}_f\cdot\boldsymbol{r}}$ the matrix element in Eq. (14) can be evaluated (*49*).

$$|\langle f|\tau|\tilde{a}\rangle|^2 \sim \cos^2\gamma\,|\Phi_a(\boldsymbol{k}_f) + r'\,\Phi_g(\boldsymbol{k}_f)\,S\,(\boldsymbol{k}_f)|^2, \tag{15}$$

where $\Phi_a(\boldsymbol{k}_f)$ and $\Phi_g(\boldsymbol{k}_f)$ are the Fourier transforms of the atomic orbitals at the adatom and the substrate atoms, respectively, $S(\boldsymbol{k}_f)$ is the bond site structure factor which determines the interference effects between electrons emitted from the various centres in the surface molecule, and $\gamma$ denotes the angle between $\boldsymbol{k}_f$ and the vector potential $\boldsymbol{A}$. $r'$ depends on the radial part of both the adatom and substrate atomic wave functions and is treated as a parameter by *Gadzuk*.

It is clear from Eq. (15) that the angular distribution depends on the angular symmetry of the adsorbate and metal orbitals involved, as well as on the bonding geometry. To illustrate the angular anisotropies to be expected, Fig. 6 shows a series of angular distributions (contours of constant intensity on a flat screen parallel to the surface) calculated by *Gadzuk* (*48*) for a $p$-type adatom orbital on a $d$-band substrate. The left hand side shows the angular distribution obtained for the adatom located symmetrically between four substrate atoms, whereas the right hand column gives the angular distribution for "bridge bonding", *i.e.* adatom symmetrically between two substrate atoms. The angular distributions are shown

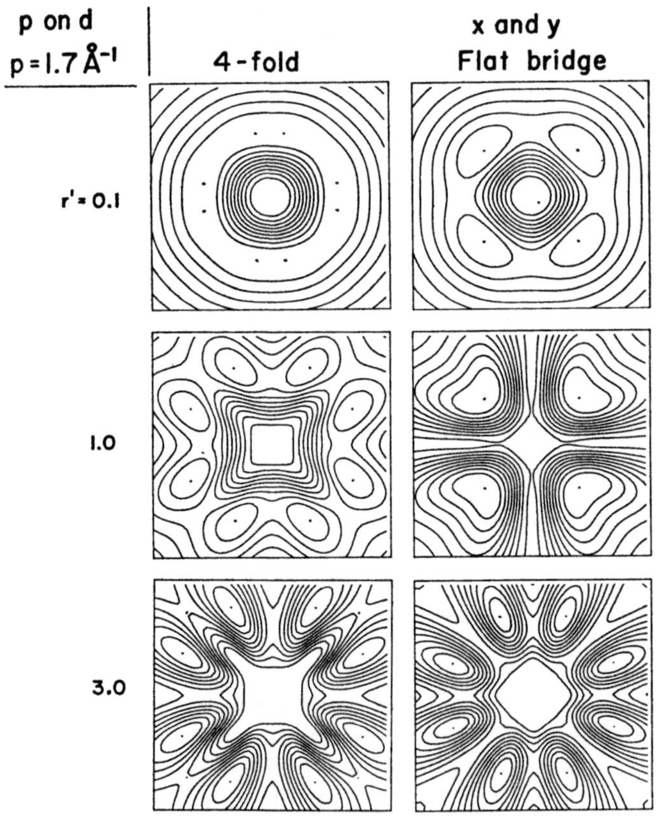


Fig. 6. Comparison between the constant intensity contours from a $p$-orbital in the flat four-fold site of a $d$-band substrate, the left column, and degenerate flat x and y bridge sites. Here unpolarised normal incidence 21.2 eV light is used. With $\varepsilon_p = -10$ eV, $p = 1.7$ Å$^{-1}$. The parameter $r' = 0.1$, 1.0 and 3.0.
[Reproduced by permission from *J. W. Gadzuk*, Solid State Comm. *15*, 1011 (1974)]

for various values of the parameter $r'$. The calculated angular anisotropy depends clearly on the bonding site geometry, but also in a pronounced manner on the parameter $r'$. It has been suggested (*49*) that measurements of the angular distribution together with calculations of this type should allow the determination of adsorbate-substrate bond geometries in a more straightforward manner than with LEED (*50—52*). As already mentioned, however, final state effects also influence the angular distribution and therefore a considerable refinement of the theory is necessary before it can be applied to real systems to determine quantitatively the bonding site geometry.

That final state scattering is of importance has been shown by *Liebsch* (*53*) and *Liebsch* and *Plummer* (*54*) who have performed calculations for a model system consisting of a localised *s*-type adsorbate orbital on a simple cubic substrate. The appropriate Bloch character of the final state wave function is taken into account within a multiple scattering theory such as has been applied in LEED calculations.

The two processes that contribute to the photoemission current are the direct emission from the adsorbate orbital into a plane wave final state $|f\rangle$ and the indirect emission from the adatom orbital via backscattering from the substrate lattice. Formally both effects are taken into account by replacing the matrix element of Eq. (14) by

$$\langle f|(1 + TG)\tau|\tilde{a}\rangle \tag{16}$$

where $T = V_L + V_L G T$ is the $T$-matrix corresponding to the lattice potential $V_L$ and $G$ is a free-electron propagator with an appropriate self-energy inserted to include inelastic effects. *Liebsch* (*53*) has evaluated the matrix element Eq. (16), approximating $|\tilde{a}\rangle$ by a localised orbital of *s*-symmetry at the adsorbate (thus neglecting those effects discussed by *Gadzuk*). Whereas the first term in Eq. (16) is simply proportional to the Fourier transform of the initial state wave function, the second term, which describes the backscattering from the surface, depends on the reciprocal lattice vectors of the substrate parallel to the surface as well as on the position of the adatom parallel and normal to the surface. As a result of the backscattering both the energy distribution as well as the angular distribution of the photoemission current from the adsorbate reflects properties of the substrate lattice.

Fig. 7 shows, for two positions of the adsorbate, the intensity as a function of final energy as calculated by *Liebsch* (*53*) for emission in the direction of the surface normal. The backscattering from the substrate is seen to cause structure in the energy distribution which is of the order of 20—50% relative to the intensity in the absence of scattering. The resonance energies are specified by the reciprocal lattice vectors of the substrate, whereas the relative intensities of these extrema depend on the bonding geometry. As demonstrated by *Liebsch* and *Plummer* (*54*) a resonance peak due to emission from a broadened adsorbate level might be strongly distorted by final state scattering, with the peak shape depending on the final state energy, the angle of detection and the electron escape depth. The larger the intrinsic half width, the more the peak maximum can be shifted, and even a splitting of the peak might occur.

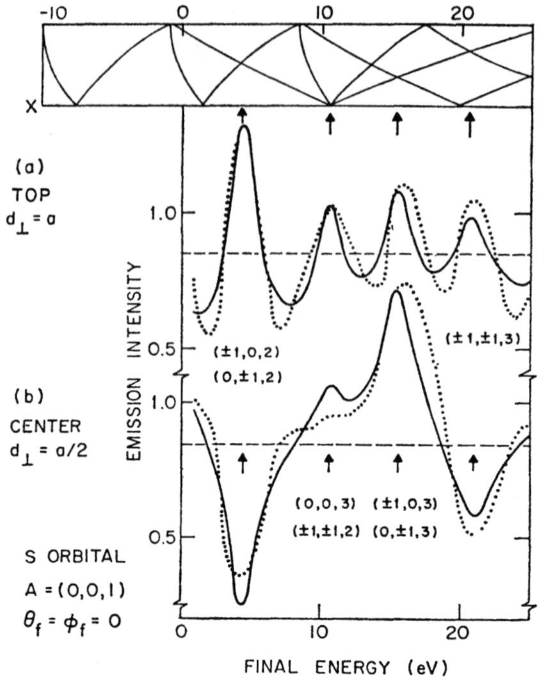


Fig. 7. Intensity (arbitrary units) as a function of final electron energy for emission from an *s* orbital adsorbed (a) in the position (0, 0, *a*) and (b) in the position (*a*/2, *a*/2, *a*/2), relative to the substrate. Single scattering (solid curves), multiple scattering (dotted curves), and no scattering (dashed curves). The arrows indicate the resonance energies specified by the reciprocal lattice vectors, and the corresponding band crossings in the free-electron band structure at the top of the figure.
[Reproduced by permission from *A. Liebsch*, Phys. Rev. Letters *32*, 1203 (1974)]

A remarkable feature in Fig. 7 is that the single scattering intensity, corresponding to the replacement of $T$ by $V_L$ in Eq. (16), is seen to be a good approximation to the multiple scattering intensity. This is in contrast to the experience with LEED calculations where extensive multiple scattering calculations are necessary to obtain meaningful results (*50—52*). It has been argued (*53*) that the more favourable situation encountered here is due to the fact that the adatom acts as a spherical source in contrast to the incoming plane wave in LEED.

Final state scattering effects also influence the angular distribution of the photoelectrons. As shown by *Liebsch*, maxima and minima are produced in the photoemission intensity as a function of the final polar angle. Their position depends on the substrate symmetry, whereas the relative intensities depend on the adsorbate bonding geometry. These modulations of the intensity are again of the order of 20—50% of the intensity without scattering. Finally, modifications of the intensity as a function of the azimuthal angle, which reflect the symmetry of the substrate lattice, are expected.

Thus there is evidence now that final state scattering effects are of importance for the understanding of UPS spectra from adsorbate covered surfaces. Angular anisotropies are expected to result from the coherent backscattering of the photoemitted electrons from the substrate lattice. Also the energy distribution of the photoelectrons might be influenced.

To sum up the present state of the theory, there have been several important steps forward in the elucidation of the photoemission process on adsorbate covered surfaces. The importance of interference effects between the photoelectrons from the adsorbate and the substrate has been pointed out by *Penn*. The dependence of the angular distribution on the symmetry of the initial state has been evaluated by *Gadzuk*. *Liebsch* and *Plummer* stressed the influence of final state scattering by the substrate lattice on both energy and angular distribution of the photoemitted electrons. To arrive at a theory which allows the quantitative description of photoemission from real adsorbate-metal systems all these ideas have to be incorporated simultaneously and probably additional effects accounted for, *e.g.* the final state scattering by the adsorbate potential. Until the completion of such a theory it is only possible to make superficial interpretations of the large quantity of experimental data becoming available.

## 4. Some Experimental Results

### 4.1. The CO/Nickel System

The first paper to show that adsorbate energy levels could be detected in UPS was published three years ago by *Eastman* and *Cashion* (*5*). On polycrystalline nickel they showed that chemisorbed CO exhibits peaks in the photoelectron energy distribution at 7.5 and 10.7 eV below $E_f$ (see Fig. 8). They tentatively identified the —7.5 eV peak with the $5\sigma$ molecular orbital on CO and the − 10.7 eV peak with the $1\pi$ orbital. More recent spectra shown in Fig. 9 due to William's group show similar results for nickel single crystal surfaces: the position of the energy levels appears remarkably insensitive to the crystal face (see also Table 1). The first conclusions to be drawn are simply that in agreement with theory adsorbate levels are observed in practice and show the expected broadness. In addition *Penn* (*39*) has used the difference spectra based on the Eastman and Cashion results for CO/nickel to demonstrate the presence of his "antiresonance" on the low binding energy side of the −7.5 eV adsorbate level. However the often observed complexity of the difference spectrum from metals like tungsten suggests that such a simple interpretation is by no means generally applicable.

Because the CO/nickel system was one of the first investigated and because the adsorbate resonances are energetically separated from the metal *d*-band, several groups have developed models for the bonding mechanism based on the assumption that the adsorbate spectral function is revealed directly by the photoelectron spectrum. This has led to a proposed re-assignment in the CO/nickel spectra, and indeed in other spectra of CO adsorption systems, as we see in the following.

All quantitative calculations done so far to interpret the energy levels in UPS of CO chemisorbed on nickel are based on molecular orbital theory. These methods have had considerable success in explaining UP-spectra of free molecules. It is however by no means obvious that an equivalent treatment of the chemisorption problem can be regarded as having the same power in explaining the UP-spectra from adsorbate layers.

Four, rather different types of calculation have been performed. The first is due to *Doyen* and *Ertl* (*55*), who use an Anderson model including overlap and consider in their treatment only the coupling of the $2\pi^*$ and $5\sigma$ orbitals of CO to the metal *d*-band. As we shall see below, all other approaches which do not have these restrictions predict the inclusion of the $1\pi$ and $4\sigma$ orbitals of CO and of the *s*-band of nickel to be essential for an understanding of the bonding. Thus it is unlikely that such a calculation can be used to explain the CO/nickel spectrum until the other orbitals are included.

*Blyholder* (*56*) has performed a CNDO calculation on CO chemisorbed on a (111) plane group of ten Ni atoms. His main results are summarised in the following. Firstly he finds that five orbitals of $Ni_{10}$—CO can be assigned to the highest five occupied molecular orbitals of gaseous CO. The effect of chemisorption is to reduce the energy gap between the $1\tilde{\pi}$ and $5\tilde{\sigma}$ orbitals and to shift the $5\tilde{\sigma}$ and $4\tilde{\sigma}$ orbitals to lower energy. (The $1\tilde{\pi}$ orbital of chemisorbed CO corresponds to the $1\pi$ orbital etc.) The bonding of the carbon atom to the Ni atom largely involves nickel *s* and *p* orbitals with little contribution from the *d* orbitals. According to *Blyholder* the first peak in the UP-spectrum which is due to CO contains both the $1\tilde{\pi}$ and $5\tilde{\sigma}$ orbitals and the second peak is the $4\tilde{\sigma}$ orbital. The calculated gap between the peaks is about 6.5 eV compared to the experimental value of about 3 eV. Before discussing the additional methods it should be mentioned that CNDO calculations of Ni atom clusters (*57*) seem to have some relevance in predicting electronic properties.

The self-consistent statistical exchange ($X_\alpha$) multiple scattered wave method, which is well described in Ref. (*58*), has also been used by *Batra* and *Robaux* (*59*) to assign the peaks in the UP-spectrum of CO/Ni.

Having recognized that the muffin-tin approximation, widely used in molecular calculations, is rather severe, they use the overlapping atomic spheres concept (*60*). This concept has been considered to lead to an improved description of ionisation potentials of molecules where a substantial fraction of the charge due to the valence electrons is distributed over the interatomic region of constant potential (*60*). The SCF $X_\alpha$ calculations on $Ni_5$—CO and $Ni_4$—CO clusters yield three main peaks, one due to the *d*-band and two which can be related to the $1\pi$ and $5\sigma$ orbitals of CO. The relative energies of the peaks agree well with the experimental values. In contradiction to the initial tentative assumption of *Eastman* and *Cashion* and to the CNDO results of *Blyholder*, the first peak of the adsorbed CO is found to be due to the $1\tilde{\pi}$ orbital and the second due to the $5\tilde{\sigma}$ orbital.

Because of the latter contradiction an *ab initio* Hartree-Fock calculation has been performed (*61*). To reduce the computer time only a linear Ni—CO molecule has been studied. Such a calculation cannot be assumed to give accurate ionisation potentials of a chemisorbed system, but if the surface molecule picture is valid, it may supply us with some relevant information. The calculation predicts similar to the CNDO calculation that the first "CO peak" is due to the $1\tilde{\pi}$ and $5\tilde{\sigma}$ orbitals and the second peak due to the $4\tilde{\sigma}$ orbital of CO. The ordering of the $1\tilde{\pi}$ and $5\tilde{\sigma}$ orbitals of chemisorbed CO is, however, different from the ordering

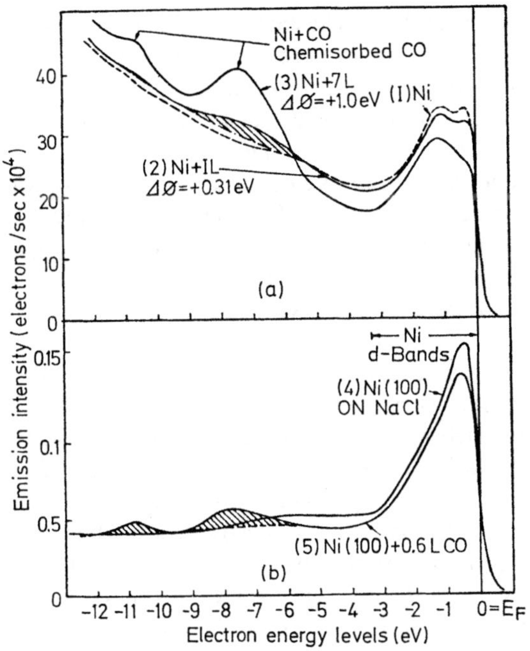


Fig. 8. Eastman and Cashion's spectra from the CO/nickel system. (a) Spectra at $h\nu = 21.2$ eV for (1) Ni, (2) Ni + 1 L exposure of CO and (3) Ni + 7 L of CO (1 L = 1 Langmuir = $1 \times 10^{-6}$ Torr sec). (b) Spectra at $h\nu = 40.8$ eV for (4) Ni and (5) Ni + 0.6 L of CO.
[Reproduced by permission from Phys. Rev. Letters *27*, 1520 (1971)]

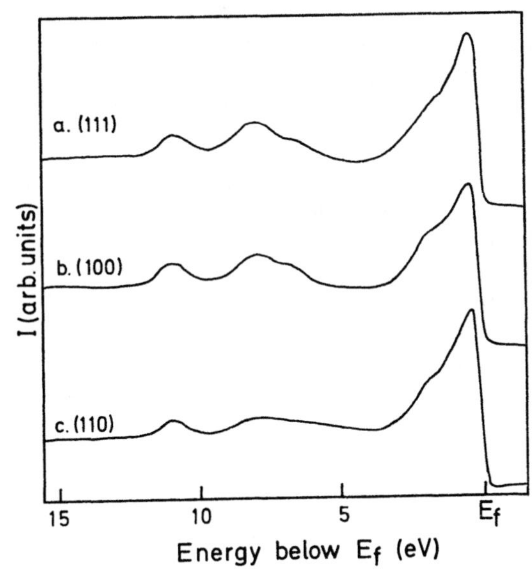


Fig. 9. UP-spectra from CO-saturated nickel surfaces at $h\nu = 40.8$ eV and room temperature. (a) Ni(111) (b) Ni(100) and (c) Ni(110). (*P. M. Williams*, private communication)

predicted by CNDO. It might be the case that this is due to an insufficient number of Ni atoms being included. In any case both calculations are not rigorous enough to solve this problem of assignment *inside* the first peak. The gap between the two main peaks is calculated to be about 4.5 eV and is in better agreement with experiment than the corresponding CNDO value.

In Table 1 the energies of the five orbitals $1\tilde{\pi}_{1,2}$, $5\tilde{\sigma}$, $4\tilde{\sigma}$ and $3\tilde{\sigma}$ of chemisorbed CO and the corresponding energies of CO gas are shown for CNDO, Hartree-Fock and SCF $X_\alpha$ (*Batra* and *Robaux* give the $1\tilde{\pi}$ and $5\tilde{\sigma}$ energies only by a figure). As we have seen in Fig. 9 there is certainly evidence for a doublet structure in the first peak, thus supporting the CNDO and *ab initio* SCF calculations. *Eastman* and *Demuth* (*63*) have recently investigated this doublet nature in somewhat more detail on the nickel (111) surface. They performed crude angular distribution measurements and saw clear evidence for two structures in the first level at −6.2 and −7.5 eV, with relative amplitudes depending on the angle of emission. Also supporting this assignment are some recent results of *Fuggle* and *Menzel* (*64*) on the CO/ruthenium system. (As will be seen below the degree of conformity amongst systems where CO is non-dissociatively adsorbed leads to the conclusion that an essentially similar bonding mechanism is valid throughout). Their results show that neither with HeII nor with $MgK_\alpha$ is the $4\sigma$ level observed in the expected ("gas-phase") position that would be consistent with the earlier assignment of *Eastman* and *Cashion* (*5*). Furthermore, at the $MgK_\alpha$ photon energy, the second peak is more intense than the first: in the gas phase the $4\sigma$ line is the strongest of the three at this energy [see Table 2 and Ref. (*65*)]. The $X_\alpha$ assignment has been critizised by *Brundle* (*66*), who pointed out that the *relative* change of the intensity ratio $I_{5\sigma}/I_{4\sigma}$ in the gas phase on going from HeI to HeII would not be compatible (see Table 2). The ratio of intensity of peak 1 to intensity of peak 2 always becomes smaller on going from HeI to HeII. Such a problem does not however arise with the CNDO or *ab initio* SCF assignments. Further good evidence supporting the latter comes from the metal carbonyls (*67*). *Hillier et al.* (*68*) have calculated (*ab initio* SCF and $\Delta$SCF methods) the electronic structure of $Ni(CO)_4$ and found that the $5\tilde{\sigma}$ and $1\tilde{\pi}$ orbitals are very close together.

Table 1. The one-particle energies of free and chemisorbed CO as obtained from the CNDO and *ab initio* (with equivalent basis sets for free and chemisorbed CO) and from the SCF $X_\alpha$ method[a]). All energies in au

| Orbital | Orbital energy CNDO | *ab initio* | Orbital | Orbital energy CNDO | *ab initio* | SCF $X_\alpha$ |
|---|---|---|---|---|---|---|
| $5\tilde{\sigma}$ | −0.714 | −0.714 | $5\sigma$ | −0.638 | −0.553 | −0.533 |
| $1\tilde{\pi}_1$ | −0.756 | −0.702 | $1\pi$ | −0.762 | −0.652 | −0.614 |
| $1\tilde{\pi}_2$ | −0.774 | −0.702 | | | | |
| $4\tilde{\sigma}$ | −0.993 | −0.872 | $4\sigma$ | −0.903 | −0.795 | −0.761 |
| $3\tilde{\sigma}$ | −1.599 | −1.622 | $3\sigma$ | −1.649 | −1.554 | −1.356 |

[a]) No numbers were given in Ref. (*59*) for CO/nickel.

Table 2. Relative intensities in gas phase photoelectron spectra of CO [after *Huang et al.* (*65*)]

| | IP vert. (eV) | Relative intensities HeI | HeII | Mg $K_\alpha$ |
|---|---|---|---|---|
| $5\sigma$ | 14.0 | 1.00 | 1.00 | 1.00 |
| $1\pi$ | 16.5 | 0.44 | 1.42 | 0.5 |
| $4\sigma$ | 19.7 | 0.11 | 0.63 | 2.5 |

As already mentioned, we are not yet certain whether the molecular orbital approach is accurate enough to interpret UP-spectra from chemisorbed systems. There are many properties, as for example line broadening, which cannot be explained by such a model. It seems however that assignment problems can be tackled with success.

### 4.2. CO-adsorption Systems in General and the Work Function Problem

When we look at the position of the adsorbate energy levels for CO in a variety of metals, we are struck immediately by the many similarities (see Table 3). The resonance at lower binding energy, which we refer to in the table as peak 1, lies invariably between $-7$ and $-9$ eV and that at higher binding energy (peak 2) between $-10$ and $-12$ eV. The separation ranges from 2.9 to 3.5 eV, the exception being the 2.4 eV separation measured by *Page* and *Williams* for polycrystalline platinum (*71*). There is practically no difference for example in the position of the peaks in the CO/nickel system compared to the CO/copper system, the latter being an extremely weak adsorption: at room temperature several Torr CO are required to reach the monolayer stage. When we consider this table in detail we are rapidly lead to the conclusion that the mode of bonding in the adsorption systems where CO is non-dissociatively bound to the surface is essentially similar. Thus the bonding schemes discussed in Section 4.1 appear to be of general applicability and the CNDO or *ab initio* SCF assignment to be generally favoured for the whole series of results.

That the problem is often not quite so simple as represented above is shown by adsorption on tungsten surfaces. For purposes of illustration we show in Fig. 10 and 11 clean surface and difference spectra for the CO/W (110) system at both HeI and HeII photon energies. There are distinct features associated with the species first adsorbed at lower coverages — $\beta$-CO — and then peaks due to $\alpha$-CO at higher coverages (*79*). The photoelectron results show clearly that $\alpha$-CO exhibits features (not so prominent in HeI as in HeII) at 7.9 and 11.0 eV below $E_f$, in common with other CO adsorption systems where non-dissociative adsorption occurs. $\beta$-CO on the other hand shows no peak in the $-10$ to $-11$ eV range and the difference spectra resemble more those from tungsten/oxygen adsorption systems (see below). *Bradshaw et al.* (*79*) concluded on the basis of

Table 3. Peak positions for CO on various metal surfaces

| Surface | Peak 1 (eV, rel. to $E_F$) | Peak 2 (eV, rel. to $E_F$) | Photon source | Ref. |
|---|---|---|---|---|
| Ni (poly) | 7.5 | 10.7 | He I/II | *Eastman* and *Cashion* (*5*) |
| Ni (poly) | 7.4<br>7.8 | 10.7<br>10.8 | He I<br>He II | *Joyner* and *Roberts* (*69*) |
| Ni (poly) | 7.7<br>8.0 | 11.0<br>11.0 | He I<br>He II | *Page et al.* (*72*) |
| Ni (100) | 7.8 | 11.1 | He I | *Becker* and *Hagstrum* (*70*) |
| Ni (111) | 7.6 | 10.7 | He I | *Eastman and Demuth* (*63*) |
| Ni (100) | 8.0 | 11.0 | He II | *Page* and *Williams* (*62, 71*) |
| Ni (111) | 8.0 | 11.0 | He II | *Page* and *Williams* (*62, 71*) |
| Ni (110) | 8.0 | 11.2 | He II | *Page* and *Williams* (*62, 71*) |
| Ru (100) | 7.6 | 10.7 | He II | *Madey et al.* (*73*) |
| Pt (poly) | ~8.7 | 11.7 | He I | *Biloen* and *Holscher* (*74*) |
| Pt (poly) | 9.2 | 11.6 | He II | *Page* and *Williams* (*71*) |
| Pd (poly) | 7.9 | 10.8 | He II | *Küppers et al.* (*75*) |
| Pd (111) | 7.9 | 10.8 | He I/II | *Conrad et al.* (*76*) |
| Pd (110) | 7.9 | 10.8 | He I/II | *Conrad et al.* (*76*) |
| Cu (poly) | 7.5<br>7.7 | 10.6<br>10.8 | He I<br>He II | *Joyner* and *Roberts* (*69*) |
| Mo (poly) "$\gamma$" | 7.8 | 11.2 | He I | *Atkinson et al.* (*77*) |
| Mo (poly) ("virgin") | 6.7 | 10.2 | He I | *Atkinson et al.* (*77*) |
| W (100) "$\alpha$-" | 9.0 | 12.0 | He II | *Bradshaw* and *Menzel* (*78*) |
| W (110) "$\alpha$-" | 7.9 | 11.0 | He II | *Bradshaw et al.* (*79*) |

these results that the adsorption of $\beta$-CO on W (110) involved a strong perturbation of the carbon-oxygen bond, indicating perhaps a dissociative adsorption.

As soon as we wish to compare the binding energies of the levels in such a compilation as Table 3 with each other, or indeed with the binding energies of CO in the gas phase we come up against the reference level problem. More important, if a UP-spectrum is calculated based on a finite cluster approach, an experimentally accessible reference level must be located. In the gas phase it is clear that the vacuum level provides the appropriate reference. For an atom or molecule

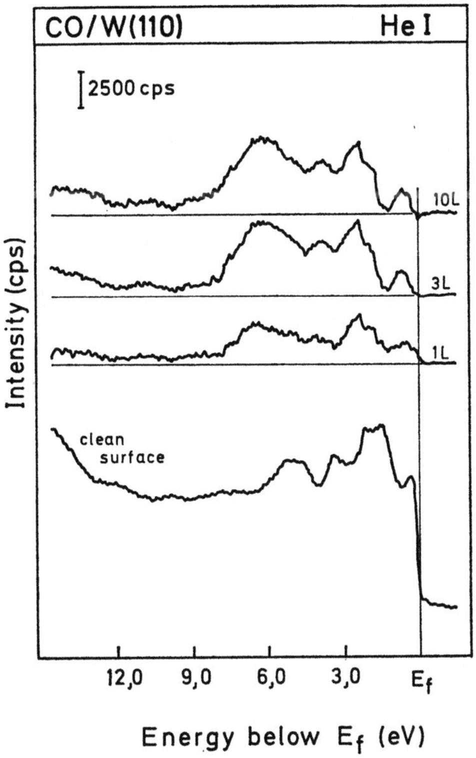


Fig. 10. UP-spectra from the CO/W(110) adsorption system at $h\nu = 21.2$ eV and room temperature. Below: clean surface spectrum and top: difference spectra after exposures of 1 L, 3 L and 10 L

on a surface, where we measure relative to $E_f$, it cannot simply be assumed that this is the case as well. Would we for instance add the clean surface work function or the work function of the adsorbate-covered surface to the binding energy relative to $E_f$?

If the latter, then as likely as not the figure added will vary according to the coverage. In order that the binding energy relative to the vacuum level does not vary appreciably there should be an approximately equal and opposite variation of the binding energy with respect to $E_f$ as a function of coverage. Put another way, if we accept that there is a unique binding energy referred to the Fermi level and independent of coverage (which is not always the case, incidentally), then there is a range of binding energies referred to the vacuum level. The problem is complicated by the fact that the adsorbed molecule sits effectively within the region associated with the surface dipole contribution to the work function and thus in a region where the crystal potential is changing rapidly. (To the average inner potential term in the work function must be added the contribution of the surface dipole, which accounts for the fact that the electron gas "spills out" beyond the atom cores. Its value will depend on the orientation of the surface, and indeed on whether there is adsorbate present). Two extreme cases can be distinguished. Firstly we have the situation whereby an absorbed atom lies

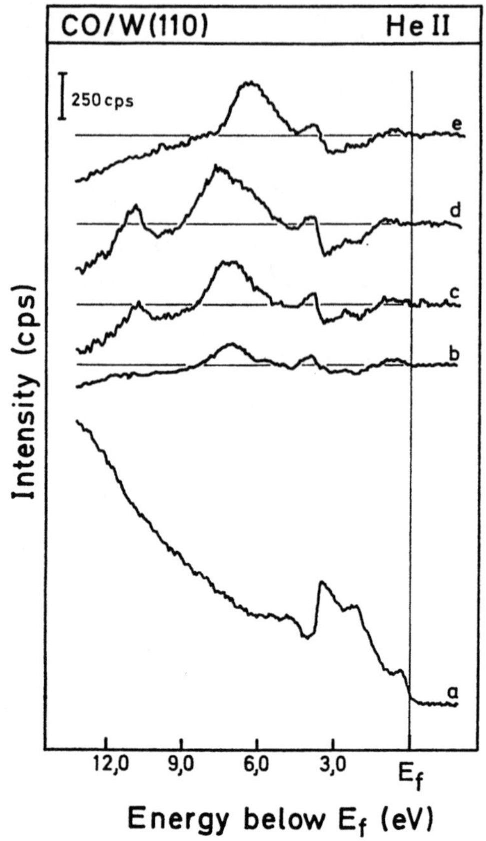


Fig. 11. UP-spectra from the CO/W(110) adsorption system at $h\nu = 40.8$ eV (a) Clean surface; (b) difference spectrum after exposure of 1 L CO; (c) difference spectrum after 10 L CO; (d) difference spectrum with surface in equilibrium with $10^{-6}$ Torr CO; (e) difference spectrum after pumping CO out and heating to approximately 600 K.
[Reproduced by permission from Chem. Phys. Letters *28*, 516 (1974)]

underneath the surface *i.e.* effectively in the "bulk". Under these conditions the obvious reference level is the average inner potential of the solid. The experimental inaccessibility of this level notwithstanding, *Citrin* and *Hamman* (*80*) have calculated the binding energies of core levels in rare gas atoms implanted in noble metals using the free atom values as starting points. The free atom binding energy was corrected in each case for the initial state chemical shift and for the final state relaxation energy due to the changed atomic environment and set equal to the binding energy referenced to the average inner potential. Calculation and addition of the surface dipole contribution of the work function gave binding energies relative to the vacuum level in tolerable agreement with experiment. As yet there are no comparable investigations in UPS on absorbed as opposed to adsorbed gases.

The other extreme is the case of adsorption outside the surface dipole layer, whereby the adsorbate interacts only weakly with the surface. A case in point appears to be α-CO on tungsten [see Ref. (*78*) and (*87*) and Table 3], where it appears as though the vacuum level might possibly be used as an effective reference level. After the formation of the strongly, most likely dissociatively bound β-CO, the α-CO species is observed at higher exposures as was mentioned above. In Table 3 we see the binding energies relative to $E_f$ for α-CO on the W(100) and W(110) surfaces. Both sets of values were obtained at high α-coverages (in equilibrium with $10^{-6}$ Torr CO) and with HeII radiation (*78*). We note here that the binding energies on W(100) are shifted higher by approximately 1 eV relative to those from the W(110) system, but that the peak separation remains constant. The nature and strength of the bonding is expected to be the same in each case and there are probably comparable relaxation contributions. The constant separation of the peaks lends some support to this view. The observed difference in the binding energies relative to $E_f$ could therefore be put down to the difference in energy between the Fermi level and the "modified" vacuum level (characterised by the work function and work function change due to β-CO). Perhaps it is too dangerous to go this far, but we do note that for the clean (100) surface the generally accepted work function value is 4.5 eV (*82*) and for the (110) surface there are several values, one of which is as high as 5.4 eV (*82*). Unfortunately there is little data for the CO-covered surfaces but large differences between the systems are not expected, and we already have a possible 0.9 eV difference in the right direction.

The more usual situation of the adsorbate in the surface dipole region is as yet unaccounted for. In fact we must refer to a "new" average inner potential, displaced from the one considered in the implanted atom experiment, when we wish to make comparison with the gas. By the same token it may be said that the photoelectrons emitted from an adsorbate atom or molecule experience a "modified" surface dipole (*83*). It is difficult enough to calculate the surface dipole contribution for a clean metal surface and so even more difficult to calculate the *modified* surface dipole contribution. Thus although we might have success in calculating orbital energies with finite cluster models, for example, and in making due allowance for relaxation and correlation effects, we still cannot calculate peak positions relative to an experimentally accessible reference level.

### 4.3. Adsorbed Hydrocarbons

*Eastman* and *Demuth* (*63, 84*) have published UP-spectra for adsorbed hydrocarbons in two interesting papers which seek a method to distinguish relaxation shifts from bonding shifts. Some of their data is presented diagrammatically in Fig. 12. In the case of benzene, for example, (Fig. 12a) the position of the gas phase levels are given in the top row. In the second and third rows the levels for several layers of condensed benzene and for chemisorbed benzene are given respectively. By comparing weakly-bound condensed benzene with gaseous benzene it was noted that all relative orbital energies and intensities were unchanged, but there was a reduction in all binding energies relative to the vacuum level of ~1.4 eV, having taken into account the modified work function. The energy shift was ascribed to relaxation effects. In comparing chemisorbed benzene with condensed benzene, the σ-orbital binding energies are further reduced due

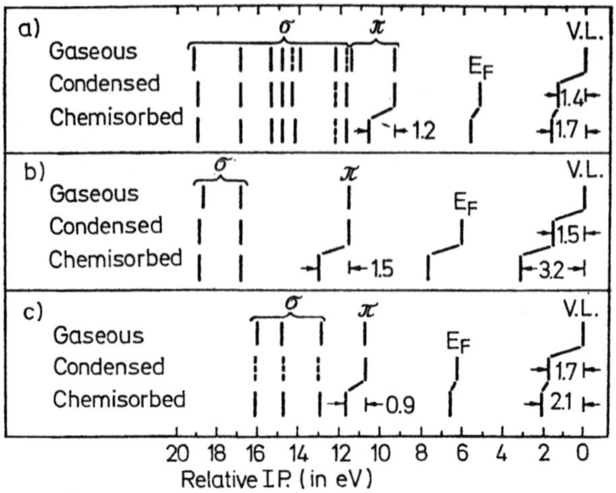


Fig. 12. Vertical ionisation energies, Fermi ($E$.) and vacuum levels (V. L.) for gaseous (1), condensed and chemisorbed phases of (a) benzene, (b) acetylene and (c) ethylene, all plotted relative to $\sigma$-orbital ionisation potentials (I. P.) for the gas phase. Relaxation shifts are given by the vacuum level shifts while bonding shifts are given by relevant $\pi$-orbital shifts. [Reproduced with permission from *J. E. Demuth* and *D. E. Eastman*, Phys. Rev. Letters *32*, 1123 (1974)]

to an increased relaxation shift, whereas the binding energy of the two uppermost doubly degenerate $\pi$-orbitals is increased due to a bonding shift. That the interaction between unsaturated hydrocarbons and metal surfaces is due primarily to $\pi$-bonding effects has been postulated for many years. The extent of the $\pi$-bonding shift was calculated as follows. All the $\sigma$-levels experience the same reduction in binding energy relative to the gas phase vacuum level, independent of whether they are condensed or chemisorbed. It is then assumed that the $\sigma$- and $\pi$-relaxation shifts from the gas phase values are equal in magnitude enabling the $\pi$-bonding shift to be calculated; for benzene it is 1.2 eV. For ethylene (Fig. 12c) a value of 0.9 eV is obtained ,which could be compared with the $X_\alpha$ results of *Rösch* and *Rhodin* (*85*). The latter have taken two "clusters" containing each a nickel-diatom in order to simulate two possible modes of bonding to the surface. In the complex representing the $\sigma$-diadsorbed ethylene the Ni—Ni bond was taken as parallel to the plane of the ethylene molecule, and in the $\pi$-complex the Ni—Ni bond was arranged perpendicular to that plane and symmetric to the C—C bond. In both complexes the ethylene remains to a great extent unchanged as can be seen from the charge distributions of the orbitals in Figs. 2 and 3 of Ref. (*85*). Also consistent with the photoelectron spectra is the prediction that only the $\pi$-level should interact strongly. The calculated $\pi$-bonding shift was 0.72 eV for the $\pi$-complex and 0.30 eV for the di-$\sigma$-complex. The $\pi$-complex configuration was then favoured by the authors, although as they themselves point out, a more realistic "cluster" would contain more nickel atoms. *Brundle* (*66*) contends that the experimental value of 0.9 eV is open to question because of the assumption of identical relaxation shifts for $\sigma$- and $\pi$-orbitals. *Eastman* and *Demuth* noted that such an assumption would not be true generally but that in this case (a) the relaxation shifts of both $\pi$ and various $\sigma$ orbitals seem to be identical for the condensed species and that (b) chemisorption bonding is observed to be a weak perturbation on the molecule. The problem of estimating the size of the relaxation shift is an additional one; underneath we still have the question of how we should adjust energy scales to bring gas phase spectra into correspondence with those from the adsorbed state *i.e.* the reference level problem.

On of the more fascinating results of *Demuth* and *Eastman* (*84*) is described in Fig. 13. The chemical state of ethylene adsorbed on nickel surfaces has long been of interest. Although earlier workers found evidence for acetylenic species, *Morrow* and *Sheppard* (*86*) were unable to detect the characteristic C—H vibrations in the IR spectrum of silica-supported $C_2H_4$/Ni. In Figs. 13a and 13b difference spectra for chemisorbed acetylene and ethylene respectively on Ni(111) at 100 °K are depicted. On warming to 230 °K the difference spectrum for chemisorbed ethylene changes dramatically giving a spectrum identical to that from acetylene. Thus chemisorbed ethylene undergoes at this temperature dehydrogenation to chemisorbed acetylene. An acetylenic species was also obtained by room temperature adsorption of ethylene on Ni(111). The difference between these results and the IR results is probably due to the vastly different types of surfaces used in the two experiments.

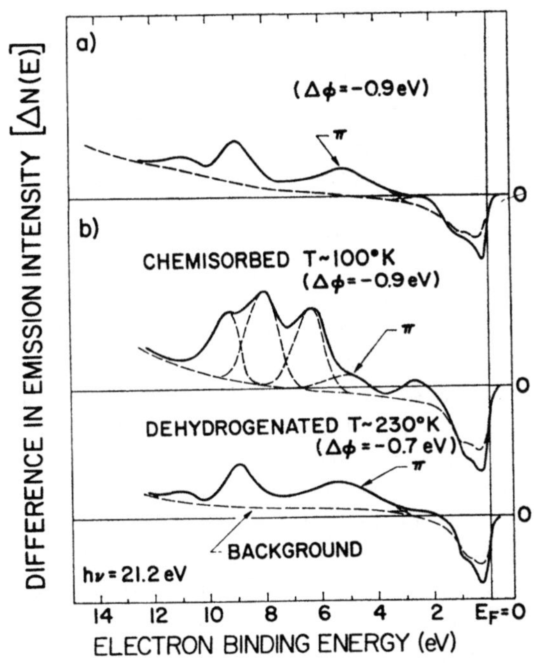


Fig. 13. UP-spectra of chemisorbed ethylene and acetylene from Demuth and Eastman. (a) Difference in emission $\Delta N(E)$ for $1.2 \times 10^{-6}$ Torr-sec exposure to acetylene at T ~ 100 or 300 K (b) $W(E)$ for chemisorbed ethylene (exposure of $1.2 \times 10^{-6}$ Torr sec at T ~ 100 K) and for dehydrogenated ethylene (obtained by warming to T ~ 230 K or with an initial exposure at T ~ 300 K).
[Reproduced by permission from Phys. Rev. Letters *32*, 1123 (1974)]

More recently *Plummer et al.* (*87*) have published a photoelectron study of the decomposition of ethylene on W(110). They came to the conclusion that at room temperature partial dehydrogenation to an acetylenic species also occurs. On heating to higher temperatures complete dehydrogenation takes place at 500—600 °K followed eventually by rupture of the C—C bonds at ~1100 °K to give usual carbon contamination in LEED. Acetylene adsorption on platinum, and benzene adsorption on nickel, copper and copper-nickel alloys have been studied by *Biloen* and *Holscher* (*74*). William's group has studied the dehydrogenation of ethylene on both polycrystalline nickel and nickel single crystals.

### 4.4. Oxygen Adsorption

Oxygen chemisorption has been studied in UPS by a number of authors (*5, 34, 72, 78, 88—94*) and in Figs. 14 to 16 we show some representative examples of this work. In Fig. 14 is shown the sequence of HeI difference spectra obtained during the adsorption of oxygen on a tungsten (100) surface (*88, 89*). Apart from the changes in the region of the W *d*-band, the two most prominent features are the growth of a positive peak between −5 and −7 eV and the loss of the surface state (*95, 41, 42*) peak at −0.5 eV. The attenuation of the latter proceeds more slowly as a function of oxygen coverage than is the case with hydrogen or CO adsorption. At exposures up to 0.3L a peak 6.0 eV below $E_f$ grew, which may be linked to the (1 × 4, 4 × 1) structure that is obtained in LEED in this exposure range (*96, 88*). The maximum in the $O^+$ ion yield curve for the $\beta_2$ state in Madey's electron induced desorption (EID) measurements (*97*) occurs at approximately 0.5L and corresponds to the maximum intensity of the (1 × 4, 4 × 1) structure. At exposures after 1L, which corresponds to the appearance of the $\beta_1$ state and a streaky (1 × 4, 4 × 1) LEED pattern, the main positive UPS peak became broader and appeared to consist of two overlapping states. By 10L this structure was about 3—4 eV wide and contained two peaks at 5.2 and 6.5 eV below $E_f$. Heating to 1500 °K produced the uppermost difference spectrum, which corresponded to a sharp p(2 × 2) LEED pattern. The observed sequence of spectra suggests that there is a single chemisorption state in the region of the (1 × 4, 4 × 1) structure. With increasing coverage this level splits into two overlapping levels. It seems probable that these two levels at 6.5 eV and 5.2 eV correspond to the $\beta_2$ and $\beta_1$ states from EID measurements, although it is difficult at this stage to say which is which. These results were correlated with XPS results taken in the same system: there was also evidence here for a second adsorption state present after about 1L. Thus in this particular case the doublet structure appears to be due to the presence of two adsorbate species on the surface and not to the final state scattering effects discussed by *Liebsch* (*53*) and *Liebsch* and *Plummer* (*54*). This particular experiment shows a good example of the type of phenomenological interpretation that can be made in the photoelectron spectroscopy of adsorbed gases. UPS is compared with other techniques, preferably *in situ* and correlations are made with the number of adsorbed species and the temperature at which they leave the surface. A similar study has been made of the CO/W(110) system (*79*).

In Fig. 15 is reproduced a HeI spectrum from *Page et al.* (*72*) for the chemisorption and oxidation of a polycrystalline nickel surface. (Here the energy scale

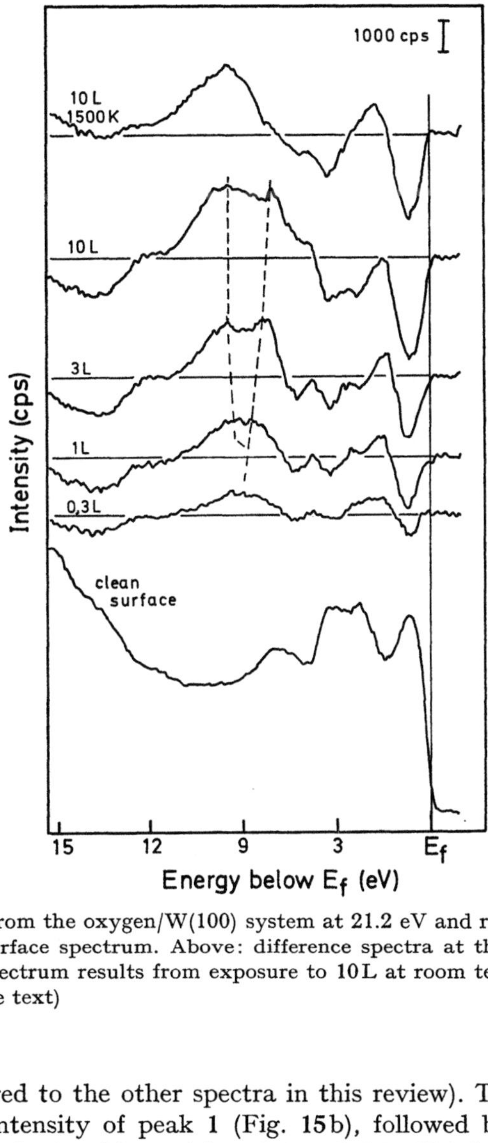


Fig. 14. UP-spectra from the oxygen/W(100) system at 21.2 eV and room temperature, after (*88*). Below: clean surface spectrum. Above: difference spectra at the indicated exposures. The top difference spectrum results from exposure to 10 L at room temperature followed by heating to 1500 K (see text)

is reversed compared to the other spectra in this review). The initial effect is a reduction in the intensity of peak 1 (Fig. 15b), followed by shifts of peaks 2 and 3 towards the final oxide position, 2.0 and 5.2 eV below $E_f$, while peak 1 decreases to near zero intensity (Fig. 15d). The sharp attenuation of peak 1 occured also in the corresponding CO, $C_2H_4$ and $H_2S$ spectra, but not in the case of $CO_2$ or $H_2O$, although the adsorbate resonances associated with the latter were also strong. The difference spectra for this system would probably reveal more characteristic details, although it would be foolish in general not to consider at the same time the real spectra in these experiments.

The third example of oxygen adsorption shown is for the oxygen/silver system in Fig. 16. Here one notes immediately the larger exposure times that are required to bring an adsorbed layer onto the surface, because of the low sticking coefficient. The difference spectra from the oxygen/Ag(110) system (*89*) are radically different

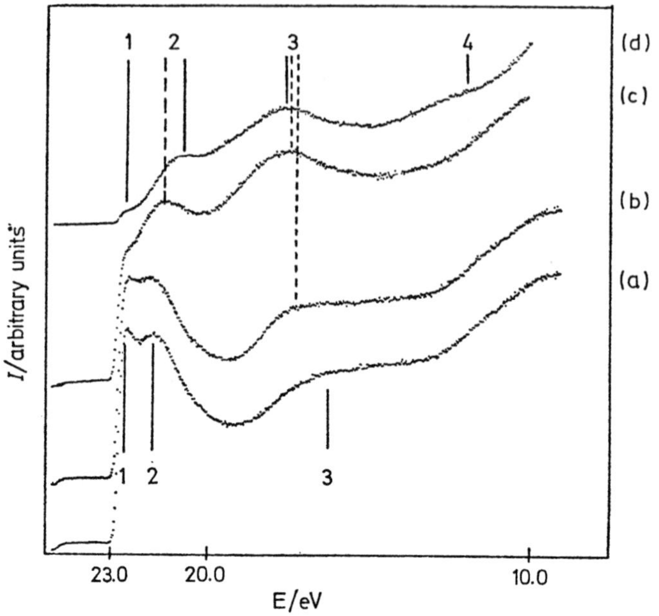


Fig. 15. UP-spectra from the oxygen/Ni (poly.) adsorption system at 21.2 eV and room temperature from *Page et al.* (a) clean nickel; (b) clean nickel + 2 L $O_2$; (c) clean nickel + 6 L $O_2$; (d) clean nickel + 56 L $O_2$ (saturation).
[Reproduced by permission from J. Chem. Soc., Faraday Trans. I *70*, 1769 (1974)]

from the corresponding tungsten or nickel spectra. The adsorbate resonances occur in the region of the *s*-band, although any small positive features below −4 eV would be masked by the strong attenuation of emission from the *d*-bands. (See Section 4.5 below). This would indicate that the interaction takes place primarily between adsorbate orbital (s)[3]) and *s* electrons from the metal. This in turn indicates a weaker chemisorption bond, which is borne out by other measurements: the heat of adsorption for oxygen on Ag(110) has been estimated at ~120 $k_J$/mole (*98*) compared with 385 $k_J$/mole for W(110) (*99*), for example. In the case of tungsten and the other transition metals it would appear that the interaction takes place mainly with the *d*-band.

## 4.5. Some General Observations

Now that we have had the chance to look at a selection of experimental results we can discuss some more general points associated with the spectra and the theory. Using the techniques described above in Section 2 we obtain UP-spectra of adsorbate-covered surfaces either as "real" spectra or as "difference" spectra. The most important change compared to the clean, uncovered surface is the observation of adsorbate "resonances" as positive peaks in the difference spectra.

[3]) Presumably oxygen 2*p*.

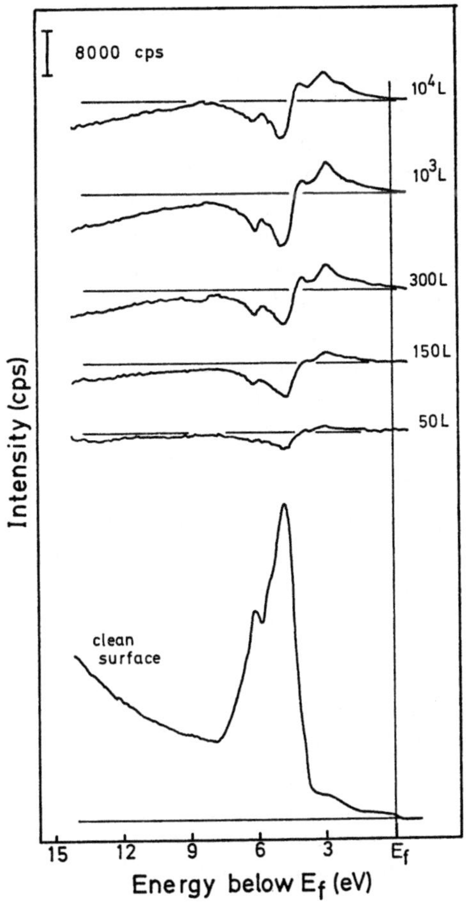


Fig. 16. UP-spectra from the oxygen/Ag(110) system at 21.2 eV and room temperature, after (*89*). Bottom: Clean surface spectrum; and above: difference spectra at the indicated exposures

This, by no means trivial result is in full accord with the theory. The fact that the resonances are generally broad is also as expected. Unfortunately the situation is often much more complicated. Firstly there can be structure in the adsorbate resonance, for example CO/Ni (*63*) or oxygen/W(100) (*88*), and secondly the effects of adsorption on the structure associated with the clean metal surface is often dramatic. The structure in the adsorbate resonance can be due to initial state effects *i.e.* two overlapping levels from the same adsorbate as we suspect for CO adsorption, or two distinct adsorption states *e.g.* $\beta_1$- and $\beta_2$-oxygen on W(100), or even structure in the spectral function itself due to correlation effects *e.g.* *Brenig* and *Schönhammer* (*38*). On the other hand we have seen in Section 3 how, according to *Liebsch*, final state scattering effects could also be responsible. As far as other changes in the clean surface spectrum are concerned, we would expect to a first approximation to see only features in the spectrum associated with the changes in the surface density of states (*91, 100, 101*) upon formation of the metal adsorbate complex. Photoemission from the bulk should be unaffected. Closer examination of the spectra, however, seems to show that attenuation effects play a role. There is obviously the possibility of an overall attenuation of emission from the unperturbed metal states of the bulk due to the decreased effective source depth after adsorption. This often accounts for the negative peaks in the difference spectra which correspond to sharp features in the metal *d*-band. The oxygen/Ag system of Fig. 16 is a good case in point (*89*). Such an effect is often, but not always (as we see from Fig. 16), accompanied by a rise in the number of inelastically scattered electrons at low kinetic energies. Another mechanism, other than the influence of bonding, which can cause structural changes in the metal *d*-band is a final state scattering effect. The photoemitted electrons from the bulk can be diffracted by the adsorbate layer when the atom or molecules forming the scattering centres assume a periodic array. Such effects can lead to either positive or negative features in the difference spectra. The explanation of negative features in terms of "antiresonances" should perhaps not be forgotten here. As we saw in Section 3 the antiresonance should accompany the main resonance as a negative feature on the low binding energy side according to the theory of *Penn* (*39*). So far it has not been possible to see from the experimental spectra whether this effect is operative: the difficulties usually lie in distinguishing it from other effects which are also present.

Two important aspects of UPS remain to be discussed. Firstly most of the work published so far has been performed with one excitation energy, namely 21.2 eV (HeI). It would be of great benefit to be able to vary the photon energy over a wide range. A start has been made by *Bradshaw et al.* (*89*), who looked at three adsorption systems using both HeI and HeII. They have observed interesting energy-dependent effects associated with the adsorbate resonance in the oxygen/W(110) system, which might be connected with the final state scattering effects of *Liebsch* (*53*). *Feuerbacher* and *Adriaens* (*91*) have also looked at several systems using both 10.6 eV and HeI radiation. In general we expect a higher excitation energy to bring a simpler clean surface spectrum: the upper state can almost be considered structureless and if we can assume that the oscillator strengths vary only slowly with energy then the band structure of the metal will be easy to elucidate. It will also be easier to separate out bulk and surface contributions in the manner of *Feuerbacher* and *Christensen* (*100, 101*). At higher photon energies (*e.g.* HeII as opposed to HeI), we also suspect that UPS is more surface sensitive (see Section 1). From the point of view of the adsorbate levels the variation in energy would assist in assignment problems as we saw in the discussion on CO in Section 4.1.

The second important aspect concerns angular distributions. In Section 3 we were able to show how the model calculation of Gadzuk predicts strong angular anisotropies in the adsorbate resonance from the metal-adsorbate-complex. The investigation of these anisotropies could lead to much more information on the geometrical structure of the chemisorption bond. As yet there are only two reports of such experiments in purpose-built apparatus.

In a preliminary account by *Waclawski et al.* (*34*) the UP-spectra at various angles of emission in one azimuth have been measured for the oxygen/W(100) system. As seen in Section 4.4 above they also see two peaks inside the main resonance at −5.2 and −6.3 eV. As a function of polar angle of emission the intensity ratios swap over as the angle probe (see Section 2 above) moves away from the normal position.

Apart from crude angular distribution measuremnets on the oxygen/W(110) system (*89*) and on the CO/nickel system (63), the only other instance is a study by *Engelhoff* and *Perry* (*102*) on the hydrogen/W(100) system, where strong angular anisotropies have been observed. Their analysis of the data at saturation coverage after the manner of *Smith et al.* (*103*) tends to indicate a two-dimensional band structure in the hydrogen-induced energy levels. The intensity of all the peaks in the difference spectra, with the exception of the one at 5 eV, drop to zero when the angle of photon incidence is zero. This supports the hypothesis of a two-dimensional band structure: the electric vector has no potential gradient to interact with at the surface as it does when it is perpendicular to the surface. The −5 eV peak is however assigned to orbitals localised around the hydrogen atoms.

During the manuscript stage of this review the authors have learnt of other accounts of UPS, which are shortly to be published. *Menzel* (*104*) in particular has reviewed recent developments in both UPS and XPS, concentrating on the usefulness of combining the methods with other surface sensitive techniques. *Plummer* (*105*) reviews UPS in some detail, drawing on much of his own as yet unpublished work for examples.

*Acknowledgements.* The authors acknowledge useful discussions with *W. Brenig, J. Fuggle, D. Menzel* and *K. Schönhammer*. They also wish to thank *I. P. Batra, G. Ertl, J. W. Gadzuk, D. L. Perry* and *E. W. Plummer* for sending preprints prior to publication.

## 5. References

1. *Turner, D. W., Baker, A. D., Baker, C., Brundle, C. R.:* Molecular photoelectron spectroscopy. New York: Wiley 1970.
2. *Richards, W. G.:* J. Mass Spectry. Ion Phys. *2*, 419 (1969).
3. *Cederbaum, S. L.:* Theoret. Chim. Acta *31*, 239 (1973).
4. *Eland, J. H. D.:* Photoelectron spectroscopy. London: Butterworth 1974.
5. *Eastman, D. E., Cashion, J. K.:* Phys. Rev. Letters *27*, 1520 (1971).
6. *Ashley, J. C., Ritchie, R. H.:* Phys. Stat. Sol (b) *62*, 253 (1974).
7. *Kleinman, L.:* Phys. Rev. B *3*, 2982 (1971).
8. *Powell, C. J.:* Surface Sci. *44*, 29 (1974).
9. *Ibach, H., Rowe, J.:* Phys. Rev., *B 10*, 710 (1974).
10. Faraday Discussion of the Chemical Society No. 58, 1975, in press.
11. *Schrieffer, J. R.:* J. Vac. Sci. Tech. *9*, 561 (1972).
12. *Gomer, R.:* Critical reviews in solid state sciences *4*, 247 (1974).
13. *Grimley, T. B.:* In: Molecular processes on solid surfaces. New York: McGraw Hill 1969.
14. *Penn, D. R.:* Surface Sci. *39*, 333 (1973).
15. *Gadzuk, J. W.:* Surface Sci. *43*, *44* (1974).
16. *Samson, J. A. R.:* Techniques of vacuum ultraviolet spectroscopy. New York: Wiley 1967.
17. *Haensel, R., Kunz, C.:* Z. Angew. Phys. *23*, 276 (1967).
18. Proceedings of the International Symposium for Synchrotron Radiation Users (ed. *Marr* and *Munro*). Daresbury Laboratory Report, DNPL/R 26, 1972.

19. *Codling, K.:* Rep. Prog. Phys. *36*, 541 (1973).
20. *Eastman, D. E., Grobman, W. D., Freeouf, J. L., Erbudak, M.:* Phys. Rev. B *9*, 3473 (1974).
21. *Wedler, G.:* Adsorption. Weinheim: Verlag Chemie 1970.
22. *Somorjai, G. A.:* Principles of surface chemistry. New York: Prentice-Hall 1972.
23. *Redhead, P. A., Hobson, J. P., Koruelson, E. V.:* The physical basis of ultrahigh vacuum. London: Chapman and Hall 1968.
24. *Tracy, J. C.:* J. Chem. Phys. *56*, 2736 (1972).
25. *Bradshaw, A. M., Menzel, D.:* Vakuum-Technik, *24*, 15 (1975).
26. *Leder, L. B., Simpson, J. A.:* Rev. Sci. Instr. *29*, 571 (1958).
27. *Johnstone, A. D.:* Rev. Sci. Inst. *43*, 1030 (1972). — *Arnow, M., Jones, D. R.:* Rev. Sci. Instr. *43*, 72 (1972).
28. *Kuyatt, C. E., Simpson, J. A.:* Rev. Sci. Instr. *38*, 103 (1967).
29. *Sar-el, H. Z.:* Rev. Sci. Instr. *38*, 1210 (1967).
30. *Sevier, K. D.:* Low energy electron spectrometry. New York: Wiley 1972.
31. *Smith, N. V., Traum, M. M.:* Phys. Rev. Letters *31*, 1247 (1973).
32. *Traum, M. M., Smith, N. V., DiSalvo, F. J.:* Phys. Rev. Letters *32*, 1241 (1974).
33. *Smith, N. V., Traum, M. M.:* Surface Sci. *45*, 745 (1974).
34. *Waclawski, B. J., Vorburger, T. V., Stein, R. J.:* J. Vac. Sci. Tech., *12*, 301 (1975).
35. *Cederbaum, S. L., Domcke, W.:* J. Chem. Phys. *60*, 2878 (1974).
36. *Newns, D. M.:* Phys. Rev. *178*, 1123 (1969).
37. *Grimley, T. B.:* J. Phys. C: Solid St. Phys. *3*, 1934 (1970).
38. *Brenig, W., Schönhammer, K.:* Z. Physik *267*, 201 (1974).
39. *Penn, D. R.:* Phys. Rev. Letters *28*, 1041 (1972).
40. *Anderson, P. W.:* Phys. Rev. *124*, 41 (1961).
41. *Waclawski, B. J., Plummer, E. W.:* Phys. Rev. Letters *29*, 783 (1972) .
42. *Feuerbacher, B., Fitton, B.:* Phys. Rev. Letters *29*, 786 (1972).
43. *Berkowitz, J. Ehrhardt, H.:* Phys. Letters *21*, 531 (1966).
44. *Carlson, T. A., McGuire, G. E.:* J. Electron Spectry. *1*, 209 (1972) and references cited therein.
45. *Ritchie, B.:* J. Chem. Phys. *61*, 3279 (1974); *61*, 3291 (1974) and references therein.
46. *Mahan, G. D.:* Phys. Rev. *B 2*, 4334 (1970).
47. *Gadzuk, J. W.:* J. Vac. Sci. Technol. *11*, 275 (1974).
48. *Gadzuk, J. W.:* Solid State Commun. *15*, 1011 (1974).
49. *Gadzuk, J. W.:* Phys. Rev. B, *10*, 5030 (1974).
50. *Andersson, S., Kasemo, B., Pendry, J. B., Van Hove, M. A.:* Phys. Rev. Letters *31*, 595 (1973).
51. *Forstmann, F., Berndt, W., Büttner, P.:* Phys. Rev. Letters *30*, 17 (1973).
52. *Duke, C. B., Lipari, N. O., Laramore, G. E., Theeten, J. B.:* Solid State Commun. *13*, 579 (1973).
53. *Liebsch, A.:* Phys. Rev. Letters *32*, 1203 (1974).
54. *Liebsch, A., Plummer, F. W.:* in Ref. (*10*).
55. *Doyen, G., Ertl, G.:* Surface Sci. *43*, 197 (1974).
56. *Blyholder, G.:* J. Vac. Sci. Technol. *11*, 865 (1974).
57. *Blyholder, G.:* Surface Sci. *42*, 249 (1974).
58. *Johnson, K. H.:* J. Chem. Phys. *45*, 3085 (1966). — *Slater, J. C., Johnson, K. H.:* Phys. Rev. *B 5*, 844 (1972). — *Johnson, K. H., Smith, F. C.,* Jr.: Phys. Rev. *35*, 831 (1972). — *Slater, J. C.:* Advan. Quantum Chem. *6*, 1 (1972). — *Johnson, K. H.:* Advan. Quantum Chem. *7*, 143 (1973).
59. *Batra, I. P., Robaux, O.:* J. Vac. Sci. Tech., *12*, 242 (1975).
60. *Rösch, N., Klemperer, W. G., Johnson, K. H.:* Chem. Phys. Letters *23*, 149 (1973).
61. *Cederbaum, L. S., Domcke, W., von Niessen, W., Brenig, W.:* Z. Physik B, in press.
62. *Williams, P.:* private communication.
63. *Eastman, D. E., Demuth, J. E.:* Jap. J. Appl. Phys., Suppl. 2, Pt. 2, 827 (1974).
64. *Fuggle, J. C., Madey, T. E., Steinkilberg, M., Menzel, D.:* Phys. Letters *51 A*, 163 (1975).
65. *Huang, J.-T. J., Ellison, F. O., Rabalais, J. W.:* J. Electr. Spectry. *3*, 339 (1974).
66. *Brundle, C. R.:* Discussion contributions, Ref. (*10*).

67. *Lloyd, D. R.:* Discussion contribution, Ref. (*10*).
68. *Hillier, I. H., Guest, M. F., Higginson, B. R., Lloyd, D. R.:* Mol Phys. *27*, 215 (1974).
69. *Joyner, R. W., Roberts, M. W.:* J. Chem. Soc., Faraday Trans. I, *70*. 1891 (1974).
70. *Becker, G. E., Hagstrum, H. D.:* J. Vac. Sci. Tech. *10*, 31 (1973).
71. *Page, P. J., Williams, P. M.:* in Ref. (*10*).
72. *Page, P. J., Trimm, D. L., Williams, P. M.:* J. Chem. Soc. Faraday Trans. I *70*, 1769 (1974).
73. *Madey, T. E., Fuggle, J., Menzel, D., Steinkilberg, M.:* to be published.
74. *Biloen, P., Holscher, A. A.:* in Ref. (*10*).
75. *Küppers, J., Conrad, H., Ertl, G., Latta, E. E.:* Jap. J. Appl. Phys., Suppl. 2, Pt. 2, 225 (1974).
76. *Conrad, H., Ertl, G., Küppers, J., Latta, E. E.:* in Ref. (*10*).
77. *Atkinson, S. K., Brundle, C. R., Roberts, M. W.:* Chem. Phys. Letters *24*, 175 (1974) and in Ref. (*10*).
78. *Bradshaw, A. M., Menzel, D.:* Ber. Bunsenges. Physik. Chem. *78*, 1140 (1974).
79. *Bradshaw, A. M., Menzel, D., Steinkilberg, M.:* Chem. Phys. Letters *28*, 516 (1974).
80. *Citrin, P. H., Hamann, D. R.:* Chem. Phys. Letters *22*, 301 (1973).
81. *Bradshaw, A. M.:* Discussion contribution in Ref. (*10*).
82. *Riviere, J. C.:* In: Solid state surface science, p. 179 (ed. Green). New York: Dekker 1969.
83. *Citrin, P. H.:* private communication.
84. *Demuth, J. E., Eastman, D. E.:* Phys. Rev. Letters *32*, 1123 (1974).
85. *Rösch, N., Rhodin, T. N.:* in Ref. (*10*).
86. *Morrow, B. A., Sheppard, N.:* Proc. Roy. Soc. (London) *A 311*, 391 (1969).
87. *Plummer, E. W., Waclawski, B. J., Vorburger, T. V.:* Chem. Phys. Letters *28*, 510 (1974).
88. *Bradshaw, A. M., Menzel, D., Steinkilberg, M.:* Jap. J. Appl. Phys., Suppl. 2, Pt. 2, 841 (1974).
89. *Bradshaw, A. M., Menzel, D., Steinkilberg, M.:* in Ref. (*10*).
90. *Baker, J. M., Eastman, D. E.:* J. Vac. Sci. Tech. *10*, 223 (1973).
91. *Feuerbacher, B., Adriaens, M. R.:* Surface Sci. *45*, 553 (1974).
92. *Hagstrum, H. D., Becker, G. E.:* Proc. Roy. Soc. (London) *A 331*, 395 (1972).
93. *Evans, S., Evans, E. L., Parry, D. E., Tricker, M. J., Walters, M. J., Thomas, J. M.:* in Ref. (*10*).
94. *Waclawski, B. J., Plummer, E. W.:* J. Vac. Sci. Tech. *10*, 292 (1973).
95. *Forstmann, F., Pendry, J. B.:* Z. Phys. *235*, 75 (1970).
96. *Estrup, P. J., Anderson, J.:* Proceedings of the 27th Physical Electronics Conference (1967), p. 47.
97. *Madey, T. E.:* Surface Sci. *33*, 355 (1972).
98. *Engelhardt, A., Menzel, D.:* unpublished results.
99. *Kohrt, C., Gomer, R.:* J. Chem. Phys. *52*, 3283 (1970).
100. *Christensen, N. E., Feuerbacher, B.:* Phys. Rev. B *10*, 2349 (1974).
101. *Feuerbacher, B., Christensen, N. E.:* Phys. Rev. B *10*, 2373 (1974).
102. *Engelhoff, W. F., Perry, D. L.:* Phys. Rev. Letters, *34*, 93 (1975).
103. *Smith, N. V., Traum, M. M., DiSalvo, F. J.:* Solid tate Commun. *15*, 211 (1974).
104. *Menzel, D.:* J. Vac. Sci. Tech. *12*, 313 (1975).
105. *Plummer, E. W.:* In: The physical basis of heterogeneous catalysis. New York: Plenum Press, in press.


Received January 28, 1975

# Structure and Bonding: Index Volume 1-24

C. K. JØRGENSEN

**Oxidations Numbers and Oxidation States**

VII, 291 pages. 1969
(Molekülverbindungen und Koordinationsverbindungen in Einzeldarstellungen)
Cloth DM 68,–; US $29.30
ISBN 3-540-04658-5

Contents: Formal Oxidation Numbers. Configurations in Atomic Spectroscopy. Characteristics of Transition Group Ions. Internal Transitions in Partly Filled Shells. Inter-Shell Transitions. Electron Transfer Spectra and Collectively Oxidized Ligands. Oxidation States in Metals and Black-Semi-Conductors. Closed-Shell Systems, Hydrides and Back-Bonding. Homopolar Bonds and Catenation. Quanticule Oxidation States. Taxological Quantum Chemistry.

H.-H. PERKAMPUS

**Wechselwirkungen von $\pi$-Elektronensystemen mit Metallhalogeniden**

64 Abbildungen. 37 Tabellen
XI, 215 Seiten. 1973
(Molekülverbindungen und Koordinationsverbindungen in Einzeldarstellungen)
Gebunden DM 78,–; US $33.60
ISBN 3-540-06318-8

Inhaltsübersicht: Einleitung und Abgrenzung. Eigenschaften der Donatoren und Acceptoren. Proton-Additions-Komplexe. $\pi$-Komplexe. $\sigma$-Komplexe. Register.

Die Umsetzung von $\pi$-Elektronen-Systemen mit Metallhalogeniden führt in Anwesenheit von Protonen zu Proton-Additionskomplexen. Sind keine Protonen vorhanden, so entstehen $\pi$- oder $\sigma$-Komplexe. Besonders ausführlich werden die binären Systeme $\pi$-Elektronen-System und Metallhalogenid behandelt. – Während das Gebiet um die Jahrhundertwende nur von bescheidenem präparativen Interesse war, ist es heute ein eigenes faszinierendes Arbeitsgebiet geworden, in dem alle modernen spektroskopischen Methoden zur Aufklärung der Struktur- und Bindungsverhältnisse angewandt werden.

Prices are subject to change without notice
Preisänderungen vorbehalten

**Springer-Verlag**
**Berlin Heidelberg New York**

GPSR Compliance

*The European Union's (EU) General Product Safety Regulation (GPSR) is a set of rules that requires consumer products to be safe and our obligations to ensure this.*

*If you have any concerns about our products, you can contact us on ProductSafety@springernature.com*

In case Publisher is established outside the EU, the EU authorized representative is:

Springer Nature Customer Service Center GmbH
Europaplatz 3
69115 Heidelberg, Germany

**Batch number: 10620881**

Printed by Printforce, the Netherlands